American Institute of Chemical Engineers

Topical Conference On Distillation 2004

Held at the 2004 AIChE Spring National Meeting

April 25-29, 2004
New Orleans, Louisiana, USA

Printed from e-media with permission by:

Curran Associates, Inc.
57 Morehouse Lane
Red Hook, NY 12571
www.proceedings.com

ISBN: 978-1-60423-479-4

Some format issues inherent in the e-media version may also appear in this print version.

American Institute of Chemical Engineers

Topical Conference on Distillation
2005

TABLE OF CONTENTS

Session 105 - Distillation/Extraction: Modeling & Processes

Chair: Ralph H Weiland
Vice Chair: Nicholas F Urbanski

Managing Risk in the Design and Startup of Small Distillation Columns

David G. R. Short

Department of Chemical Engineering
University of Delaware
Newark, DE 19716
short@che.udel.edu

Keywords: distillation, risk

Prepared for presentation at the 2004 Spring Meeting, New Orleans, LA, April 25-29

Managing Risk in the Design and Startup of Small Distillation Columns
David G. R. Short
University of Delaware

Abstract

The design and startup of all unit operations in a processing facility represent a risk that the design will meet expectations. The risk is in both the commissioning phase as well as the eventual routine operation of the unit. This report will focus on the author's experiences with small distillation columns (< 0.7m diameter) in an industry where high purity (<100 ppm total impurities) was expected. These experiences include evaluating what-if scenarios, understanding the flexibility in the system, and closely examining industry best-practices to recognize the important fundamental parameters of a good design.

Introduction

My former employer, E. I. DuPont de Nemours & Co., produces a wide variety of fluorochemical products. These go into a variety of end uses as polymers (Teflon®), refrigerants (Freon®, Suva®) and electronic gases (Zyron®). The purity requirements on the compounds are quite stringent. Most have maximum impurity levels of 100 ppm with some products calling for < 1ppm total impurities. These are products that sell for $5 to $50 per kilo or more. In addition, the processing conditions can be wide ranging with temperatures of between -50°C to 400°C and pressure range from atmospheric to 28 bars. The compounds processed in the distillation columns are usually toxic (HF), explosive (Cl_2) or both (C_2F_4). A lot of issues were dealt with in successful commissioning of the distillation columns for these processes. Most of the columns are less than 0.7 m in diameter. Column size and investment does matter on what techniques and piloting many be done. But, if a process will not start up because a column will not work properly, column size does not count as a career impact issue. A large part of the success can be attributed to managing the process risk in the design and startup of the equipment.

Process Risk

I need to be clear that with the use of the term 'process risk' in the context of this paper. It does not refer to safety hazards that are inherent in the processing of typical materials. Rather, I am referring to the risk that the column and ancillary equipment will be specified, designed, constructed, installed, and started up correctly. The book that addresses most of these issues is 'Distillation Operation' by Henry Kister, Figure 1. (1) Henry's book on distillation design (2) is important for doing the calculations correctly but the first book is essential for understanding the issues.

A partial list of potential issues is shown in Table 1. Table 2 lists suggested heuristics on how to manage the risk of a small column design.

At last year's Distillation Conference here in New Orleans, I showed a plot that is tied to an Excel Spreadsheet. This is shown in Figure 1. One of the purposes of this plot is to determine if proposed column configuration fits within current practice. Current practice is dependent on a combination of the process, company, and industry. A recommendation is to have a good tool for a first screening of potential designs. I highly recommend that the screening methodology be something that is documented, not something that relies on people's collected memories. My spreadsheet was just such an attempt. An almost but widely ignored advice is to treat the tool as a device to map the known successful designs from the unsuccessful designs for your service. Don't use this as an excuse to never visit the 'Land of the Dragons'. Just be sure you know the location of the boundaries.

If we look at Table 1, we see there are numerous risks that I have identified that are more specific to a distillation column. The rest of this paper will be focused on discussion the items in the table.

Examples of the Risk

Theoretical Stages

The first item in the table is the risk of a column not appearing to have enough theoretical stages. In a small, packed column, the most likely reason is liquid distribution. I had two methods of managing this risk. My recommendation is to use both. One method is education of the construction and process engineers and technicians. These are the people who will be installing the internals. A thorough education on how to properly install the internals and the consequences of poor installation goes a long way to making sure that the people who will operate the system feel that they willingly own it. The idea is to not make them feel that the column is something thrust on them by the central engineering group.

The other way is to study the risk of poor flow distribution. This has been wonderfully documented by Lockett and Billingham (3). My method is very similar in that I would model two or more columns in parallel. The internal flow can then be set to some distribution and the model executed. An example of acceptable management of process risk is a column that does not quite have the number of theoretical stages but has enough extra reflux capability to compensate. This will work as long as the column is constructed properly and the physical properties have been correctly identified and modeled.

There is another risk of the design being 'too good'. This can come about from being too conservative in the design. Or, you just got lucky. While this is an important issue in estimating the capital cost of the project, once the money is spent, the money has been spent. I have experienced a problem where the reflux was being controlled by a distillate purity measurement. When the controller was put into automatic, it reduced the reflux, and consequently the liquid flow, to below the lower limit of the distributors. After a while the quality would degrade and the reflux was changed. But increasing the flow would cause an abrupt change once the distributor was working properly again. To fix the control system to always maintain a minimum flow in the column was very easy, once we understood the issue. This may cost more energy, but you are making product. And the energy usage was part of the original design.

The tower design is an area where I consider vendor involvement to be crucial. I always considered my design to be preliminary pending the design from the vendor. As far as I was concerned, if I could not trust the vendor's design, I could not trust the vendor's equipment. Once both designs were made, we compared notes and a final design was made. I tried to have an environment where nothing pertinent to the design was withheld. There are several very good, but unfortunately, proprietary examples of how this cooperation led to vastly improved designs.

Heat exchange

With heat exchange, there is the risk of being too large or too small for the process. The problem with too small is obvious. But have you ever tried to operate a column with a condenser or reboiler that was too large for the service? The problem is control of the exchangers. One can get spiking in the heat flux and with the impurities in the product.

A more subtle issue can be illustrated in the simple heat transfer equation

$$Q = UA\Delta T_{lm} \qquad (1)$$

For a given duty and heat transfer coefficient, if the area is twice as large as necessary, then the driving force is halved.

With cooling water the flow rate has to decrease. I have seen cases where the flow was reduced to the point that the water was boiling. A consequence was the water treatment chemicals precipitated and fouled the exchanger. If this is the situation of a fixed tube sheet exchanger with the water on the shell side, then one has a real problem.

If this is going to be a potential problem, I have found that the best solution is to install a recirculating loop for the water. This has the effect of increasing the water inlet temperature so that the outlet temperature can be kept at an acceptable value. This is an added expense but may be the best way to keep product moving through the system.

When a steam heated reboiler area is too large, the control system will reduce the steam pressure in the exchanger so that the driving force is correct. The problem is the control system may want to reduce the pressure to a vacuum. This can be handled if one has planned for this situation. Possible solutions that I am familiar with include using condensate pumps or partially flooding the steam chest of the exchanger.

Partial flooding may also be a solution for the condenser situation except the typical small column does not have a reflux drum, is using gravity runback, and may even have a dephlegmator condenser. So the ability to use partial flooding may be limited.

Process Control

Having a good process control study of your system goes a long way to minimize operating risk of the system. My experience is that as a bare minimum, this study will confirm that the correct method is being used to control the pressure, condenser, reboiler, and product removal flows. The real issue is determining the best set of indicators to tell us if the column is performing its task. This is an issue that is solved by having all members of the project and operating team engaged in understanding the control of the system.

With small columns, a control issue is the guidelines for good control, such as holdup in column bottoms or reflux drums, may not apply. My experience is there usually is very little holdup in the condenser systems but a large holdup in the reboilers. The condensers are usually runback systems while the reboilers are kettles sized to hold the entire contents of the column when the column is shutdown.

Summary

There are many factors that are needed to ensure that a distillation column project will be successful. This paper has focused on those issues of small columns. The main advice is to keep all parties involved. If 'over the fence' engineering is used, then that is where you may find yourself if the system does not achieve the project goals.

Bibliography

1. 'Distillation Operation', Henry Z Kister, McGraw Hill, New York, 1990, ISBN:007034910X
2. 'Distillation Design', Henry Z Kister, McGraw Hill, New York, 1992, ISBN:000070349096
3. Billingham, J. F. and Lockett, M. J., Trans. IChemE, Vol 80, p373-382 (2002)

Table 1 – Process Risk Issues with Distillation Columns

Risk	Possible Cause	Management
Not enough stages	Liquid Distribution	Study effect of lower quality distribution. Use more reflux
Too many stages	Better packing efficiency than expected	Be sure control system does not 'wind-down' the column
Too large heat exchange	Ht. transfer coefficient or driving force larger than design	Condensate pump Partial Flood steam chest Over reflux system
Too small heat exchange	Ht. transfer coefficient or driving force smaller than design	Chilled water Higher pressure steam
Condenser control		Emphasize simplicity
Reboiler control		Emphasize simplicity
Column diameter		Be sure all parties agree on design parameters and methods.
Feed Composition	Different than design	Pilot reactor with column Case study effect of feed composition

Table 2 – Design Risk Management Heuristics

- Do not deviate too far from current practice.
- Solicit ideas from many sources.
- Welcome outside review.
- Understand how all the pieces combine for a good design.
- Have a good steady-state model and understand it
- Have a good unsteady-state model and understand it
- Budget for operator training
- Never attempt to take ownership of the column away from the operators

[101c]DT003

How to Compare Packings & Internals to Obtain Comparable Quotes

by Frank Rukovena, Jr.

Presented - Distillation Forum: Evaluating Vendor Claims

2004 Spring AICHE Meeting New Orleans, LA

April 25 - 29, 2004

RC

Rukovena Consulting

Topics To Be Covered

- Packing
  - Physical & Performance Parameters
- Internals
  - Things To Watch Out For
  - What to Do About Them
- Installed Configuration
- Tower Preparation

Physical Characteristics Random and Structured Packing

- Material
  - Type
  - Material Thickness
  - Allowable Working Temperature
    - Normally important for plastic-sometimes for metal
- Surface Area per Unit Volume

Physical Characteristics Random Packing

- Piece Count / Unit Volume
- Allowable Bed Depth
- Bed Compression at Process Conditions
  - Important mostly for plastic but can be concern for thin gauge random packing in deep beds.
- Bed Life Considering Plastic Creep

Physical Characteristics Structured Packing

- Number of:
  - Layers
  - Pieces per layer
  - Weight of Heaviest Pieces
  - Wall Wiper
    - Yes/No
    - Mounted On Packing Element or Separate

Performance Parameters Random & Structured Packing

- Mass Transfer Coefficient
- Height Equiv. Per Theoretical Plate, HETP
- Pressure Drop Per:
  - Unit Depth
  - Theoretical Plate
- Maximum Capacity at Design Efficiency

Tower Considerations Existing Tower

- Tower Diameter
  - Large Enough?
- Total Height Adequate
  - Tall Enough for Packing Internals
  - Vapor Inlet and Outlet
- Number of Beds
- Total Pressure Drop
  - Pressure Drop Including Internals
- Man-hours to Install Packing and Internals

Tower Preparation/Modifications Existing Tower

- Number of:
  - Tray Rings To Be Removed
  - New Rings & Ring Completion
  - Support Clips & Ledges
  - New Nozzles & Manways
  - Shell Extension
- Cleaning Before and After Installation
- Man-hours to Complete

Tower Configuration
New Tower

- Tower Dimensions
  - Diameter - Tangent to Tangent Height
- Number
  - Beds & Internals
- Manways & Nozzles
  - Size & Number
- Installation Man-Hours
- See Items On Existing Tower List

Liquid Distributor (L.D.) Considerations

- L.D. Reduce Packing Efficiency By:
  - Introducing Uneven Liquid Flow to Packing
  - Creating Liquid Entrainment
  - Fouling Resistance
- L.D. Reduce Packing Capacity By Flooding Prematurely

Liquid Distributor Considerations

- Causes of Premature Flooding
  - Liquid Entrainment
    - Above the Distributor
    - Below the Distributor
- Inadequate Trough Freeboard
- Poorly Designed Feed Device
- Fouling

L.D. – How To Prevent Premature Flooding

- Prevent Liquid Entrainment
  - Properly Place Feed Device Discharges
  - Use Vapor Riser Covers – Re-Distributor
  - Use Liquid Conductor Pipes
- Provide 25 – 50mm Freeboard
- Use Largest Orifice Possible

L.D. Items Affecting Packing Efficiency – Design

- Uneven Flow to Packing
  - Number Orifices (Drip Points) per unit area
  - Orifice Pattern
  - Too-Low Minimum head
    - Orifice to Orifice High Variation (Micro Variation)
  - High Local Liquid Velocity
    - Micro Variation
  - High Bulk Liquid Velocity
    - Large Area Variations (Macro Variation)

L.D. Items Affecting Packing Efficiency – Design

- Ease of Assembly and Leveling
- Feed Device Liquid Discharge
  - Location Relative to Liquid Orifices
  - Relative to Vapor Riser Discharges
- Gaskets - Minimize
  - Most Critical at Low Liquid Rates
- Fouling Resistance

Liquid Distributors
How To Reduce Flow Variation

- Design Deepest Min. Head Possible
  - >50 mm
  - Level With Care
  - Feed Pipe/Trough Liquid Discharge
    - Locate Between Orifices
    - Use Energy Dissipation Device - Baffles or Packing
  - Gaskets
    - Keep Number of Joints to a Minimum
    - Install Properly
  - Prevent or Minimize Fouling

Liquid Distributors How To Reduce Fouling Potential

- Liquid Strainers – Increases Fouling Resistance
  - External – Best Option
  - Internal - Central or Individual
- Use Sidewall Orifices
  - >50mm Above Trough Bottom
- Use Largest Orifice Possible > 3mm
  - Reduce Number of Orifices to Minimum ($50/M^2$)

Liquid Distributor Entrainment - Design

- Items To Be Considered
  - Vapor Riser
    - Velocity (Area) – Inlet & Outlet
    - Cover Design
  - L.D. Discharge Streams Locations Relative To:
    - Vapor Riser Inlets and Packing Hold Down
  - L.D. Height Above Packed Bed
    - Conductor Tubes

Liquid Distributor
How To Prevent Entrainment

- **First Principle**

  Never Let the Liquid Free

Liquid Distributor
How To Prevent Entrainment

- Keep Orifice Discharge Points Away from Gas Riser Inlets.
- Use Conductor Tubes To Shield Orifice Discharge Streams.
- Use Gas Riser Covers on Re-Distributors
- Eliminate Splashing
- Use High Gas Riser Open Area - >30%

Tower
Entrainment - General

- Bed Discharge Streams Relative To High Velocity Reboiler Inlets >600mm
- Use Inlet Vapor Distributor
  - High Inlet Velocity Normal When Converting Trayed Tower to Packed Tower
- Liquid Pool Level Relative to Vapor Inlet
  - >600mm Generally Good

Conclusion

- When Comparing Packing Quotes Review The Total Tower Perspective
  - Packing for:
    - Efficiency – Capacity – Pressure Drop
  - Distributors – Re-Distributors for:
    - Flow Variation – Entrainment Creation - Fouling
  - Packing Hold Downs for:
    - Interfering with Liquid Flow - Splashing

Conclusion (cont.)

- Feed Devices
  - Discharge Velocities and Locations
- Tower Arrangement
  - Diameter and Total Height
  - Vapor Inlet Parameters
- Existing Tower Modification Costs
- Preventative Measures For:
  - Fouling
  - Entrainment

Retrofit of Depropanizer and Debutanizer Packed Column

Glenn Shiveler, Sulzer Chemtech
Dr. Don Love, Consultant
Darryl Pierce, Papco Corporation

Prepared for Presentation at 2004 AIChE Spring Meeting
Distillation Forum: Tray and Packing Efficiency
Distillation: Equipment and Applications I

Retrofit of Depropanizer and Debutanizer Packed Column

Mr. Glenn Shiveler, Design/Applications Engineer
Sulzer Chemtech USA
Tulsa, Oklahoma

Mr. Darryl Pierce, President
Papco Corporation
Warren, Pennsylvania

Dr. Don L. Love
Consultant
Tulsa, Oklahoma

Abstract

A Depropanizer and Debutanizer designed for the separation of propane, butanes, and gasoline products were equipped with new internals and #0.7 Nutter Rings™ random packing. Although the efficiency of the packing was expected to generate an HETP of 1.9 to 2.2 feet, the measured efficiency was significantly less. Consequently, the towers were not making the proper product splits. In order to improve the column performance, reflux rate of both towers was increased and the feed rate was substantially reduced to prevent flooding. Even with increased reflux ratio, the desired product purity was not achieved.

After extensive testing and evaluation, it was concluded that the efficiency of the packing was low due to inadequate liquid distribution and a chimney tray was needed to separate the liquid and vapor from the feed stream for both towers. After replacing the liquid distributors and installing a chimney tray, the tower operating parameters were then adjusted to the design values. As a result, the packing efficiency improved to a HETP value of 1.5 to 2.0 feet for both distillation columns

Design and Initial Operation
The distillation columns were designed to separate the feed into HD-5 quality propane, butanes, and gasoline. In the Depropanizer column, the propane was to be separated to the top product and the butanes and heavier components flow to the bottom, with a reflux ratio of 2 to 2.5. In the Debutanizer, the iso-butane and n-butane were to be separated to the top, and the gasoline mixture of pentanes and heavier components were to flow to the bottom of the tower, with a reflux ratio of 2 to 2.5. The packing installed was a #0.7 Nutter Ring, with a design flood of 53% and 47% for the Depropanizer top and bottom sections, 64% and 58% for the Debutanizer top and bottom section, respectively.

Two recently published papers have highlighted the design methods and retrofit challenges encountered in light-ends columns equipped with trays in gas plants. [1,2] These columns are designed using modern process simulation software to optimize the number of theoretical stages by evaluating reflux ratio, and by minimizing energy utilities associated with the condenser and reboiler. Various graphics are prepared by the equipment designer to facilitate the optimization process. Published data suggested that the HETP for 1.5" pall rings in Depropanizer service should be near 3 ft, and the HETP for 1.5" pall rings in Debutanizer service should be approximately 2.3 ft. [3] During a performance test conducted at Fractionation Research Inc., the #2.5 Nutter Rings performed with an HETP of 1.6 ft at 100 PSIA pressure with the iC4/nC4 system, similar to the operating conditions of the Debutanizer for this application. [4]

After the process design basis is established, the column packing and tower internals are designed. For this application, an orifice-pan type was used. It was designed with small vapor risers and oversized holes in the distributor plate for the liquid flow. The feed did not have any chimney tray for liquid and vapor distribution, but entered the tower through a sparger type pipe distribution header located above the vapor risers of the feed distributor. Unfortunately for the end-user, the tower internals were designed and fabricated by the equipment designer and not Sulzer Chemtech. The end-user suffered several months of poor equipment performance before contacting experienced equipment designers to solve the performance problem.

The bed support grid had an open area of greater than 95%. Also, the vapor return line from the reboiler in the bottom of the bed was adequately sized with a good distribution entrance to the bottom of the bed. Therefore, there was no maldistribution of the vapor or liquid flow from the bottom of the packing beds.

The actual operation showed butanes and pentanes in the propane product and excess propane in the bottom liquid in the Depropanizer at the design reflux ratio and feed loads. In the Debutanizer, the butanes product had a significant amount of pentanes and the gasoline had butanes present at the design reflux ratio and feed loads. To improve the split of the components, the feed load was reduced and the reflux ratio was increased from 2.5 to 5.0 for the Depropanizer and from 2.5 to 5.0 for the Debutanizer. The feed rate was reduced because of the limitation to flooding in the rectifying sections with the increase in reflux ratio.

This change in operation improved the split somewhat but did not achieve the desired product purity. In fact, it appeared that flooding was occurring in the feed section of the distillation columns.

Table 1 is a comparison of the design as compared to the actual operation.

After careful testing and evaluation, it was concluded the problem was caused by the following.

1. Inadequate distribution of vapor and liquid for the reflux to the rectifying sections and the feed liquid distribution in the distribution trays for the stripping sections.
2. Inadequate separation of liquid and vapor in the feed to each tower. The liquid was believed to be mixing with the vapor from the feed distribution vapor risers, causing excess frothing and/or foaming in the mid-section of each tower. This resulted in a loss of theoretical trays in the bottom section of the rectifying section.

Liquid Distributors

Although there was sufficient number of liquid distribution openings on the distributor, the vapor riser area was less than 10% of the tower area. The liquid head above the distributor orifice holes was nearly equal to the pressure drop of vapor through the distributor. Consequently, the vapor was believed to be competing with the liquid flow through the liquid holes and causing a high frothing, turbulent action in the top of the tower and in the feed section. This action did not necessarily cause a flooding condition, but is believed to have reduced the packing efficiency (and HETP) of the upper area of the packing for each section. This was due to mal-distribution of liquid to the packing cross section, and due to back-mixing of liquid in the bottom section of packing located just above the feed distributor.

With excess packing height required for the application, this still could have resulted in a much better distribution of the components when the reflux ratio was increased and the feed rate decreased. Therefore, it is concluded that even though this could have reduced the packing efficiency, it was believed that the more significant problem was the inadequate separation of the vapor and liquid of the feed stream into the towers.

To further confirm this, the distribution tray was removed in the reflux section of the Debutanizer. Operations with testing confirmed there was no drop in efficiency and no improvement in the separation. Therefore, it appeared the major problem was due to the inadequate vapor and liquid separation of the feed stream above the feed distribution tray.

The distribution trays were replaced with a new orifice pan type distributor with the vapor flow area in excess of 15% of the tower area, and with the liquid holes evenly distributed to maintain a liquid level above the holes of approximately 4 to 6 inches at the design reflux ratio and feed loads. A new bed limiter was also added approximately 6 inches below each distribution tray.

Chimney Tray for Feed Stream

Since it was apparent the major problem was in the feed area, a chimney tray was designed and installed in the midsection to adequately separate the liquid from the vapor above the distributor. The liquid from the chimney tray flows to the deck of the distribution tray in lieu of contact with the vapor flowing from the distributor.

Figure 1 shows the overall installation of the reflux distributors, feed distribution trays, and the feed chimney trays.

Retrofit of Distribution Trays and Chimney Tray into the Towers

To avoid welding inside the towers and/or modifying the tower height, the new distributor and collector trays were designed to be installed on the original support rings. The reflux distributor had vapor risers taller than the original risers and therefore, the mist eliminator above the reflux distribution trays had to be remounted and supported above the vapor risers by support rods. In this manner, this would have slightly reduced the height of separation in the top of the tower.

Therefore, the reflux distributors were installed at a lower level than the original trays, hanging the new trays on the original support. In this manner, about one foot of packing had to be removed from the top of each of the rectifying beds. This reduced the number of theoretical trays for the rectifying section.

For the feed section, distributors were supported on the original support rings, except the chimney tray had to also be supported on the same ring to avoid any welds on the code-stamped towers. In order to accomplish this, the distributors had to be relocated along with the bed limiter approximately two feet below the original support ring. This resulted in removing approximately two feet of the packing from the top section of each stripping section. As a result, the stripping section would have less theoretical trays than expected originally.

Evaluation of Operations After Retrofit

After the new distribution trays and feed chimney trays were installed, there was a significant improvement of bed efficiency, resulting in 12 theoretical trays (HETP of 1.7-feet) for the rectifying section and 15 theoretical trays (HETP of 1.4-feet) for the stripping section of the Depropanizer. For the Debutanizer, the testing indicated there were 8 theoretical trays (HETP of 1.6-feet) for the rectifying section and 13 theoretical trays (HETP of 1.6-feet) for the stripping section.

The reflux ratio was successfully reduced to the design value, and the feed was increased to the design rates. With the lower reflux rates, the reboiler and condenser load was reduced.

Table 2 is a summary of comparison after the retrofit was completed and operations were restored.

Table 3 provides the mechanical specifications for the #0.7 Nutter Ring packing utilized in these columns. Hydraulic calculations confirm the Depropanizer packed sections are rated at 57% and 48% capacity for the rectification and stripping section. The Debutanizer packed sections are rated at 58% and 49% based on plant data following the retrofit. Table 4 provides a summary of the packing hydraulic and efficiency calculations.

Comparison to Process Simulation

The performance of the Depropanizer was compared to a process simulation. Figure 2 shows the reboiler duty versus number of rectification stages for the Depropanizer, both prior to and following the retrofit. Figure 3 shows the reflux ratio for the Depropanizer, which shows there is a good match between the design theory and plant operation. Figure 4 shows the tray-by-tray composition profile of propane, as calculated by the simulation.

Figure 5 shows the reboiler duty plot versus rectification stages for the Debutanizer, both prior to and following the retrofit. Figure 6 shows the reflux ratio for the Debutanizer. Figure 7 shows the reboiler duty for different feed locations, which confirms the feed point is within two stages of its ideal location. Figure 8 shows the tray-by-tray composition profile for key component n-butane, as calculated by the simulation.

Conclusions

Based on this case study, it was concluded that the major problem was in the distribution of the two-phase feed without a chimney tray, although the distributor retrofit did contribute to some of the improvement. Uniform liquid distribution is required in order to obtain excellent performance from high performance packing. Thus, the design of liquid distributors and collectors is paramount to the successful operation of the packed tower. From this retrofit, it is highly recommended that a chimney tray with good separation of vapor and liquid be considered for packed towers where two-phase flow is present.

Background:

Glenn Shiveler, Design/Applications Engineer, Sulzer Chemtech USA, Tulsa, Oklahoma. Bachelor of Science in Chemical Engineering from Rutgers University, and a member of AIChE, with over nineteen years of experience servicing the mass transfer industry.

Darryl Pierce, Secretary/Treasurer of Papco Inc.. BS in Petroleum Engineering, Colorado School of Mines, member of S.P.E. Professional Engineer in Louisiana and Pennsylvania with 29 years in drilling and production operations in the Gulf of Mexico, Persian Gulf, Rocky Mountains, Texas, Louisiana, Gulf Coast and the Appalachian Basin with Pennzoil, Petroleum Engineers, Inc. and currently owner of Edward Engineering.

Don L. Love, Ph. D., P.E. semi-retired consultant, Tulsa, Oklahoma, B.S., M.S. and Ph.D. in Chemical Engineering from Texas Tech University, University of Tulsa, and LaSalle University. Member of AIChE, OSPE, Engineers Society, Professional Engineer in Texas and Oklahoma, with over 40 years of design in process development and operations for Linde Corporation (formerly Pro-Quip), Love Process Engineering, Williams Refinery Resources, Amerada-Hess, Transwestern Pipe Line, and Magnolia Petroleum.

Literature References:

[1] Process simulator effective in De-ethanizer tower revamp; Oil & Gas Journal; March 25, 2002; Atul Choudhari, Pradnya Gune, S.K. Saxena; pp. 56-60.

[2] Revamp, troubleshooting optimize NGL Depropanizer operations; Oil & Gas Journal; August 25, 2003; Daryl Hanson, Ian Buttridge; pp. 88-99.

[3] Engineering Data Book, Volume I, Gas Processors Suppliers Association; Tenth Edition; 1994; Figure 19-24.

[4] FRI Topical Report #109; M. Sakata, T. Yanagi; February 15, 1990; Test Runs #16842 & #16843.

Table One

Initial Operation before the retrofit of

	Depropanizer Design	Depropanizer Actual	Debutanizer Design	Debutanizer Actual
Diameter of Tower, ID	23.25"		15.25"	
Packing Height				
Rectification, feet	23'		16'	
Stripping, feet	23'		23'	
Reflux Ratio	3.0	5.0	2.0	5.0
Theoretical Trays	25	9	20	8
Feed, lb/hr	3,033	2,250	1,139	824
Feed Composition, mole%				
Ethane	0.649	0.676	0.94	8.886
Propane	69.510	69.790	24.656	23.849
Iso-Butane	8.669	8.770	47.498	41.887
n-Butane	13.749	13.250	11.453	11.184
Iso-Pentane	3.306	3.320	6.082	6.446
Hexane Plus	4.117	4.194	9.371	7.748
Overhead, lb/hr	1,894	1,426	744	594
Overhead Product, mole%				
Ethane	0.97	0.959		
Propane	98.53	95.264	1.287	11.448
Iso-Butane	0.43	2.463	33.747	30.676
n-Butane	0.07	1.272	64.766	53.312
Iso-Pentane		0.031	0.19	3.88
Hexane Plus		0.011	0.01	0.684
Bottoms, lb/hr	1,139	824	397	230
Bottoms Product, mole%				
Ethane	0.94	8.886		
Propane	24.656	23.849		
Iso-Butane	47.5	41.887	0.02	0.181
n-Butane	11.45	11.184	0.71	2.279
Iso-Pentane	6.082	6.446	41.973	36.506
Hexane Plus	9.371	7.748	57.3	61.034
Reboiler Duty, MBTU/hr	705	1,048	403	648
Condenser Duty, MBTU/hr	707	1,056	230	150

Table Two
Comparison of Design to Retrofit Operation

	Depropanizer Design	Depropanizer Retrofit	Debutanizer Design	Debutanizer Retrofit
Diameter of Tower, ID	23.25"		15.25"	
Packing Height				
Rectification, feet	23'	23'	16'	16'
Stripping, feet	23'	21'	23'	21'
Reflux Ratio	3.0	2.5	2.0	2.5
Theoretical Trays	25	27	20	21
HETP – Top Section, feet	1.92	1.92	2.00	2.00
HETP – Btm Section, feet	1.77	1.50	1.92	1.62
Feed, lb/hr	3,033	3,786	1,139	1,417
Feed Composition, mole%				
Ethane	0.649	0.430	0.94	0.43
Propane	69.510	70.199	24.656	27.99
Iso-Butane	8.669	8.439	47.498	46.60
n-Butane	13.749	13.630	11.453	11.22
Iso-Pentane	3.306	3.278	6.082	3.03
Hexane Plus	4.117	4.024	9.371	10.74
Overhead, lb/hr	1,894	2,369	744	970
Overhead Product, mole%				
Ethane	0.97	0.61		
Propane	98.53	99.01	1.29	0.57
Iso-Butane	0.43	0.37	33.75	37.21
n-Butane	0.07	0.02	64.77	61.89
Iso-Pentane			0.19	0.32
Hexane Plus			0.01	0.01
Bottoms, lb/hr	1,139	1,417	397	447
Bottome Product, mole%				
Ethane	0.94	0.43		
Propane	24.66	27.99		
Iso-Butane	47.50	46.60	0.02	0.00
n-Butane	11.45	11.22	0.71	0.17
Iso-Pentane	6.08	3.03	41.97	44.30
Hexane Plus	9.37	10.74	57.30	55.53
Reboiler Duty, MBTU/hr	705	1,097	188	403
Condenser Duty, MBTU/hr	707	1,042	230	473

Table Three

Mechanical Specifications for #0.7 Nutter Rings

Piece Count	4,740	pieces per cu.ft.
Specific Surface Area	69	sq.ft. per cu.ft.
Dry Packing Weight	11	pounds per cu.ft.
Void Fraction	97.8%	
Packing Factor	39	

Table Four

Summary of Packing Hydraulic Calculations

	Depropanizer		
	Design	**Initial**	**Retrofit**
Capacity, Rectification Section	53%	58%	57%
Capacity, Stripping Section	47%	54%	48%
Column Pressure Drop, Inches H2O	9	11	10
Rectification Section HETP, Inches	23		21
Stripping Section HETP, Inches	21		17

	Debutanizer		
	Design	**Initial**	**Retrofit**
Capacity, Rectification Section	64%	64%	58%
Capacity, Stripping Section	58%	59%	49%
Column Pressure Drop, Inches H2O	9	9	7
Rectification Section HETP, Inches	24		19
Stripping Section HETP, Inches	23		19

Figure One
Tower Internals

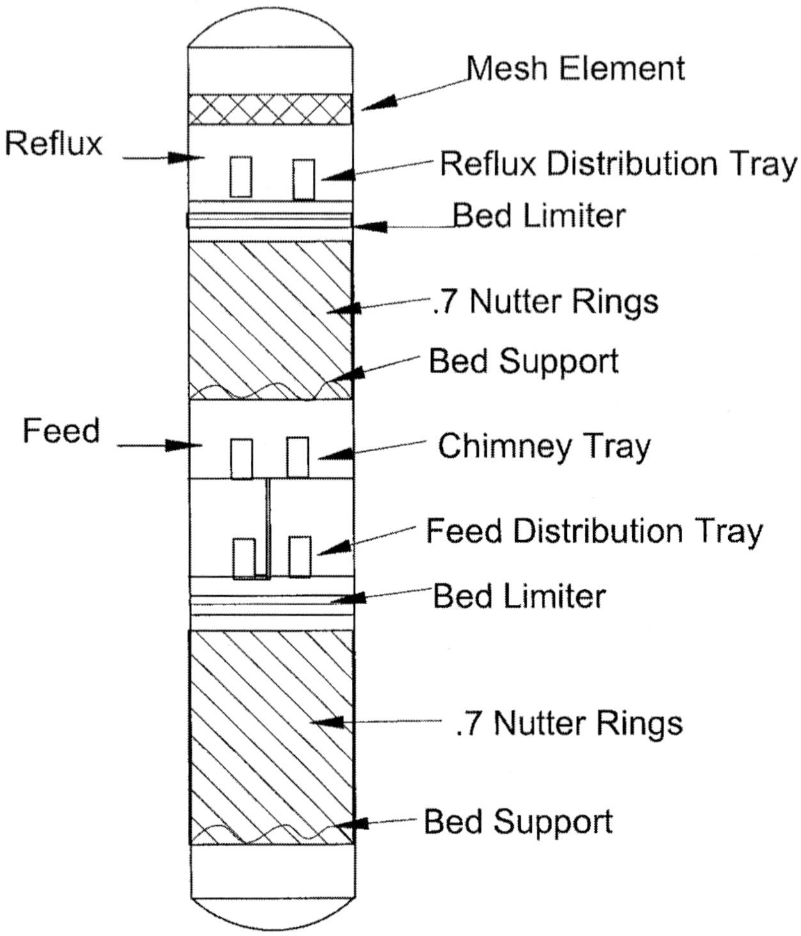

Figure 2 - Depropanizer - Reboiler Duty vs. Stages

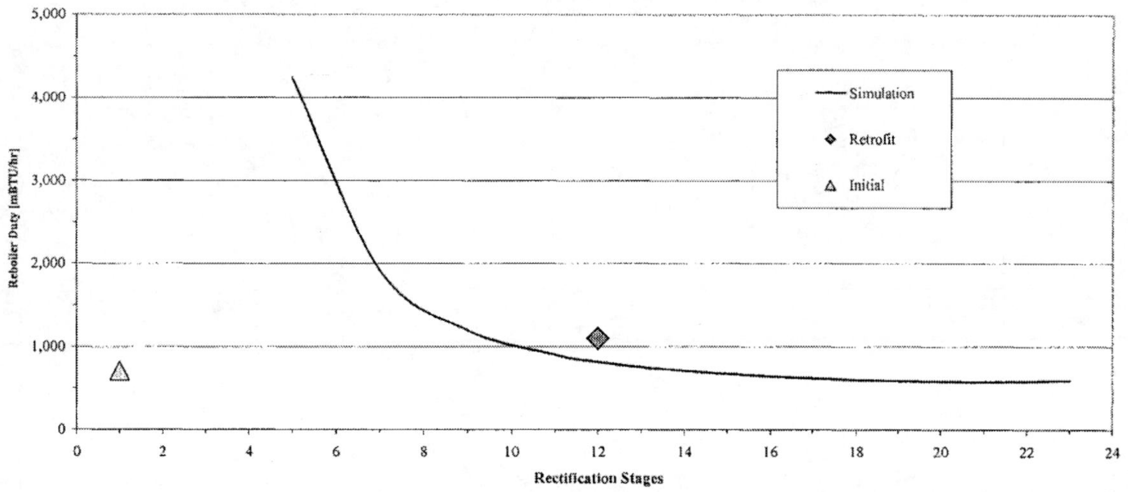

Figure 3 - Depropanizer - Reflux Ratio vs. Stages

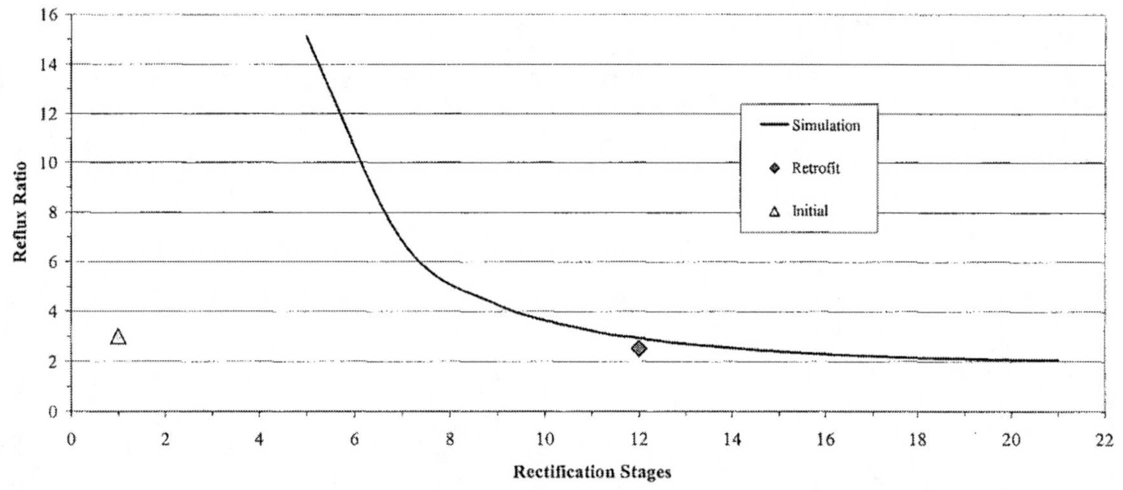

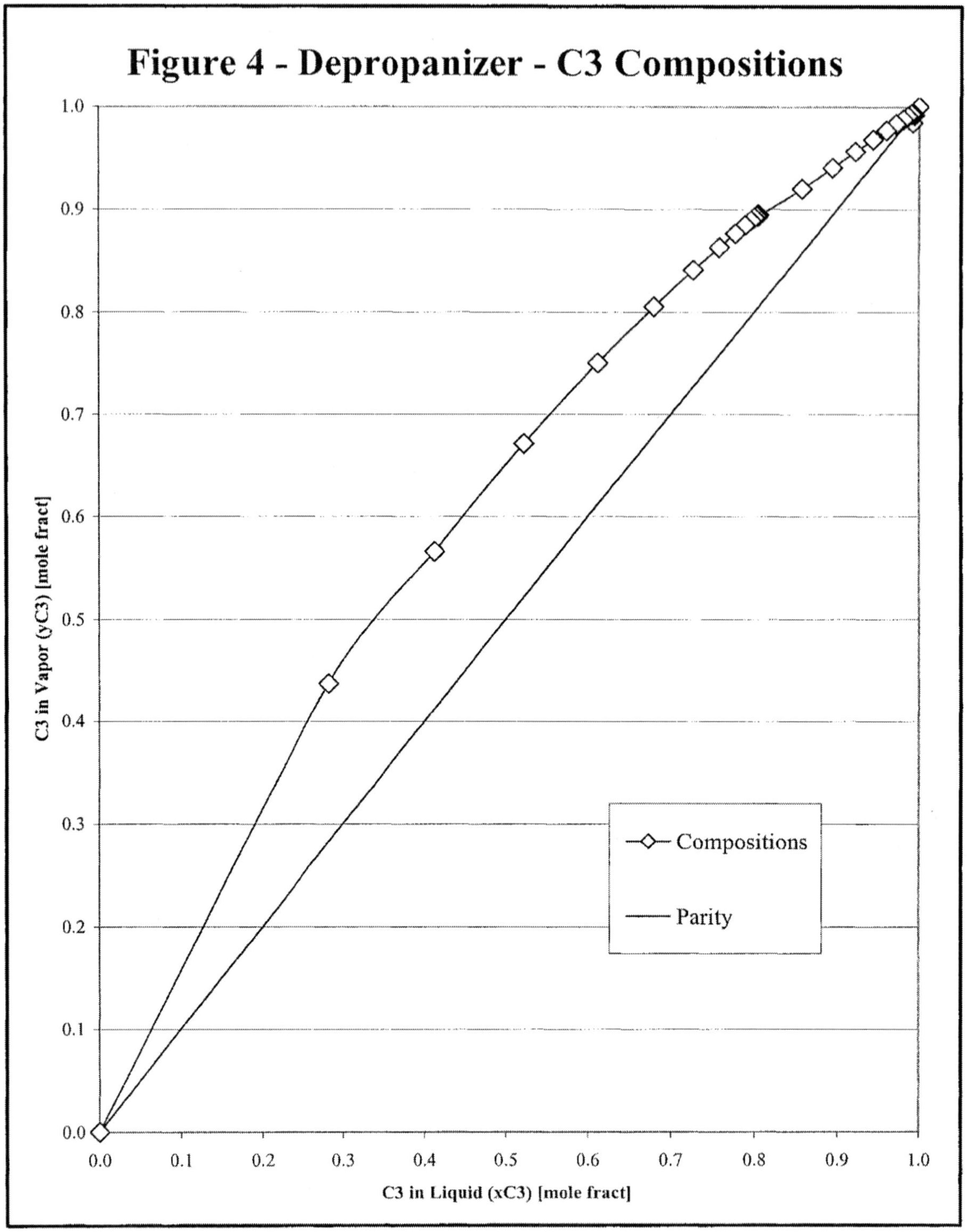
Figure 4 - Depropanizer - C3 Compositions
1.0
0.9
0.8
0.7
0.6
0.5
0.4
0.3
0.2
0.1
0.0
C3 in Vapor (yC3) [mole fract]
0.0
0.1
0.2
0.3
0.4
0.5
0.6
0.7
0.8
0.9
1.0
C3 in Liquid (xC3) [mole fract]
Compositions
Parity

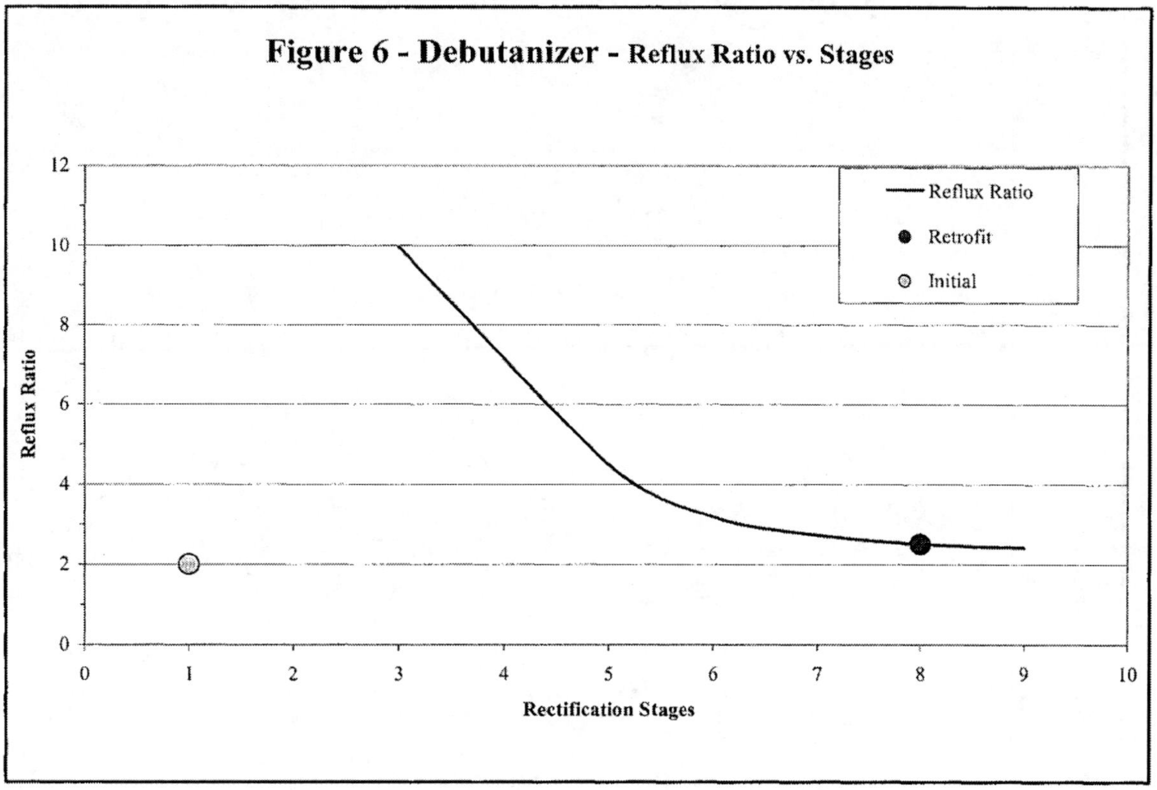
Figure 6 - Debutanizer - Reflux Ratio vs. Stages
Reflux Ratio
Retrofit
Initial
12
10
8
6
4
2
0
0
1
2
3
4
5
6
7
8
9
10
Reflux Ratio
Rectification Stages

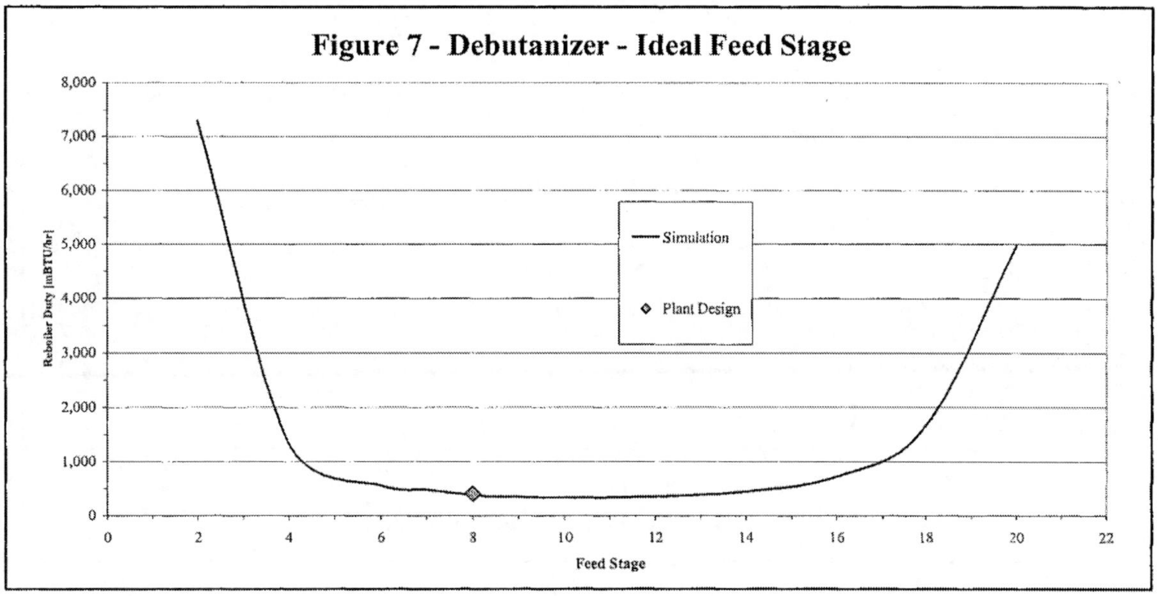
Figure 7 - Debutanizer - Ideal Feed Stage
Simulation
Plant Design
8,000
7,000
6,000
5,000
4,000
3,000
2,000
1,000
0
0
2
4
6
8
10
12
14
16
18
20
22
Reboiler Duty [mBTU/hr]
Feed Stage

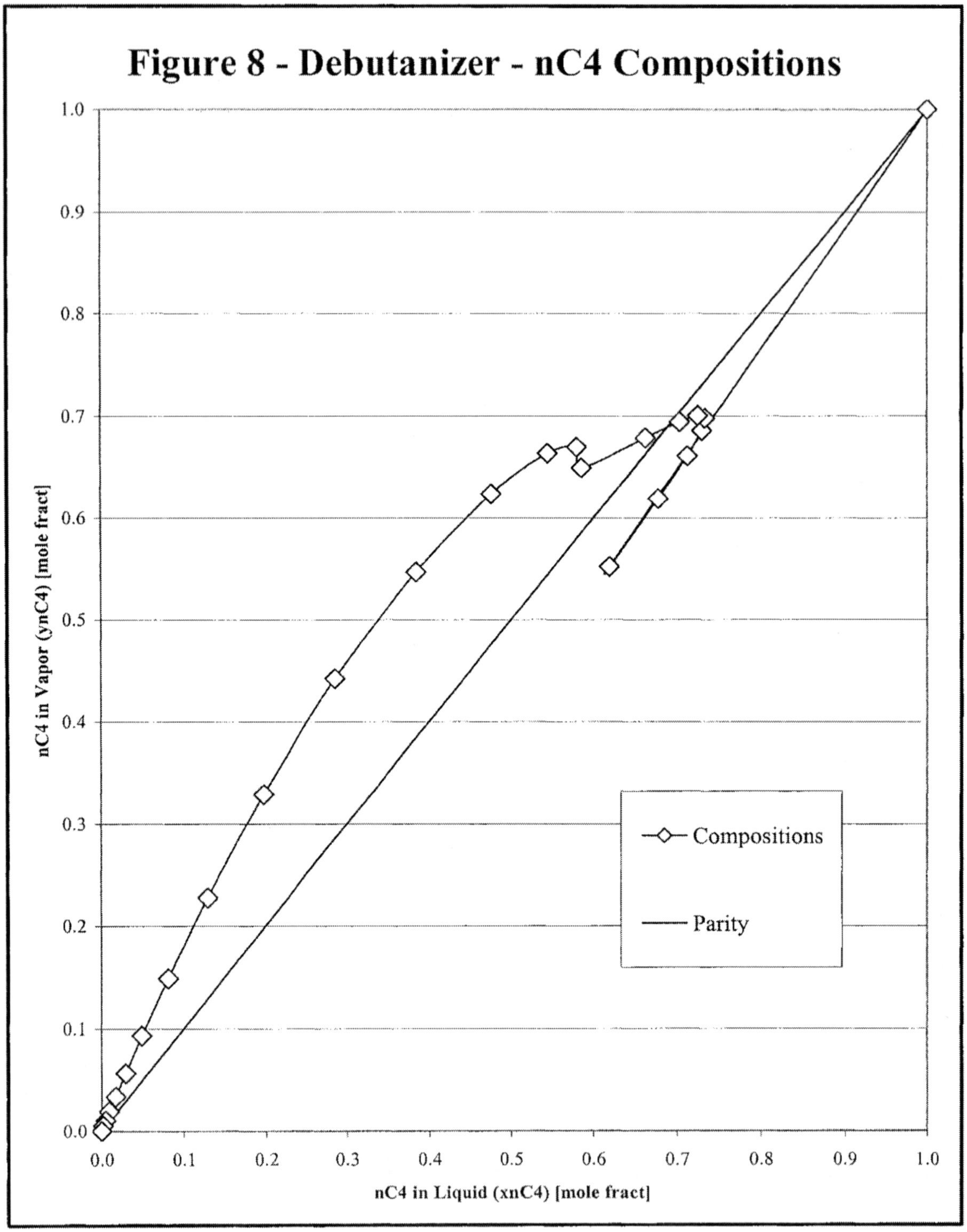
Figure 8 - Debutanizer - nC4 Compositions
1.0
0.9
0.8
0.7
0.6
0.5
0.4
0.3
0.2
0.1
0.0
nC4 in Vapor (ynC4) [mole fract]
0.0
0.1
0.2
0.3
0.4
0.5
0.6
0.7
0.8
0.9
1.0
nC4 in Liquid (xnC4) [mole fract]
Compositions
Parity

Paper No. 102c

The Recurring Jet Fuel Problem (and what it took to solve it).

Gunter Schnaibel and Maria Elizabeth P. L. Marsiglia

PETROBRAS S.A.

Keywords: column internals, trays, packing, jet fuel, kerosene, revamps

Prepared for Presentation at the 2004 AIChE Spring National Meeting, April 25 - 29, Hyatt Regency, New Orleans, LA, Distillation Session TD004: Equipment and Applications I

April, 2004

Unpublished

Abstract:

In 1997 REPLAN, Petrobras' biggest refinery carried out a major turnaround of its Crude and Vacuum Unit U-200A. The unit as whole was revamped. The crude tower was the single piece of equipment most impacted. As part of the strategy to increase unit capacity without changes to furnaces, keeping costs down, most trays in the crude tower were replaced with packing and a new bottom side draw was added.

After start-up, the jet fuel yield decreased to half of its original value with trays. The fractionation between jet fuel and diesel had deteriorated noticeably making it impossible to recover more jet fuel.

What's worse, the same problem happened to two other Petrobras refineries that went on stream or underwent revamps in 1997 and 1998. What did they have in common? They had packing in the jet fuel section. The peak of the problem was reached when REPLAN's second Crude and Vacuum Unit U-200 was revamped in 1999. Despite all preventive steps taken, it performed just as bad as unit U-200A in regard to jet fuel yield.

This paper will present the troubleshooting techniques that a group of Petrobras employees conducted to evaluate the U-200A problem: operational tests, simulations, hydraulic calculations, gamma scan surveys, talks with knowledgeable distillation experts, and where each of these steps took us. Then we will examine what was proposed to avoid the same problem in the revamp of the other unit, U-200, and why it failed. To prove that we were right in our reasoning (and that troubleshooters should not give up easily), we will finally describe the 2003 revamp of U-200A, which just started up last July, absolutely successful in solving the jet fuel problem, yielding jet fuel above required, while still sticking to packing.

Lessons learned will focus on the correct outlines for any revamp together with correct installation of packing and its associated internals. The correction of U-200 will probably be our next story... we do have a diagnosis and two possible solutions, but we have to wait for its next turnaround!

Contents:

1- Background

Petrobras' biggest refinery is Replan, located in Southeastern Brazil, with a nominal charge of 352.000 bpd (56.000 m3/d). Replan has two almost identical distillation units, U-200 and U-200A. Both performed similarly up to 1997. Their turnarounds are set at least two years apart.

During its 1997 turnaround, unit U-200A underwent a major revamp to increase its charge from 24.000 to 28.000 m3/d in maximum diesel run mode, based on a heavy crude slate. For light crudes and for jet fuel or solvent runs, the charge would remain at 24.000 m3/d.

The U-200A revamp was very inexpensive and highly attractive, presenting an internal return rate of 220%. To cut down cost and schedule, the atmospheric and vacuum furnaces were not modified. The strategy was to change the atmospheric tower internals in order to reduce pressure drop and surmount atmospheric furnace restriction. All its trays were replaced by structured packing, except the two pump around sections and the bottom stripping trays. A new bottom side draw was added, to overcome vacuum furnace limitation.

After start-up, the tower performed well in diesel runs, but for jet fuel runs, a surprise happened. It became very difficult to meet the distillation endpoint (≤ 225°C) and flash point (≥ 38°C) specs of the jet fuel stream, no matter the charge rate. The yield was down to 50% of the original with trays. The former valve trays approached 100% of the theoretical jet fuel yield. The refinery studied the problem, tried different operational schemes, but nothing really solved it. U-200A was at best getting 60% of the jet fuel yield easily achieved by U-200, which continued operating with trays, while running the same crude. Solvent production in U-200A became unfeasible after the revamp.

U-200 was under way for a similar revamp in 1999 and there were fears of duplicating the jet fuel problem. To tackle this situation, propose solutions to fix U-200A and review the design of U-200, Petrobras Headquarters nominated a team of Petrobras employees. This team is acknowledged at the end of the paper.

2- Problem Analysis

Initially, the refinery suspected of simulation errors, supposing that real liquid and vapor loads were higher than simulated, leading to an overload in the jet fuel section. Later we shall prove that actual causes were incorrect revamp outlines aggravated by undersized internals. At any rate, the refinery studies, assuming flooding, were pointing towards an incorrect direction, like increasing top pressure and increasing bottom pump around heat in expense of top pump around heat.
When our "task force" first met, after reviewing all previous work by Replan, several possible causes to the problem were put forward:

- Contamination of jet fuel due to leakage in the jet fuel sidestripper reboiler.

Hypothesis previously discarded by Replan, comparing samples of jet fuel from the crude column draw off and from the jet fuel stripper outlet.

- Contamination of jet fuel with another product, externally to the tower.

Hypothesis previously discarded by Replan, comparing samples of jet fuel from the crude column draw off and from the unit battery limit.

- Lack of theoretical stages and/or insufficient internal reflux;
    - Due to purchasing an insufficient number of theoretical stages of packing;
    - Due to packing underperformance, whether caused by liquid and/or vapor maldistribution or by unsuitable packing.

Hypothesis of lack of stages reinforced by initial simulations comparing the number of theoretical stages in unit U-200 with trays, U-200A with packing and the packing datasheet.

- Insufficient reflux.

If there really was a lack of stages, the reflux rate necessary to achieve the expected fractionation must be higher than originally simulated.

- Overloaded beds due to too low pressure;
    - Excessive entrainment;
    - Bed flooding.

Hypothesis previously discarded by Cenpes (Petrobras Research and Development Center) by evaluating the operational conditions of the beds.

- - Operational trouble with the jet fuel sidestripper;

Hypothesis reinforced by observations in the field:

- Noise close to this side stripper's bottom;
- Bubbling inside the corresponding level gauge;
- Operation of the jet fuel sidestripper feed valve with a partially opened bypass, suggesting possible flow hindrance between the crude tower and the sidestripper.

To isolate the probable causes from the unfeasible ones, a test composed of several steps was planned, each step attempting to isolate one of the effects.

3- Unit Test

The test was conducted in jet fuel run mode at 24.000 m3/d, according to design basis, processing a mixture of Merey (heavy crude) and Algerian Condensate.

The packing vendor for the three top beds was scheduled to attend the test, but they attended just part of one morning of the test.

The main idea of this test was to change only one variable at a time, while keeping everything else as constant as possible. Product yields were kept fixed except when a step required some change. Column pressure profile should also have been constant. This was not possible for all steps, since the overhead condensers operate at their limit, hindering pressure control whenever the system is overloaded. This occurred from step 5 onwards, but more seriously in the final steps, due to a progressively lighter charge, indicating the tank was heterogeneous. These were the main test steps:

Step 1- Typical unit operation.

Step 3- Increase bottom pump around to unload the top section and minimize any entrainment or flood problems. Actually, step 3 was intended to convince refinery groups that this was not the correct direction. Step 3 confirmed the column was below flood point, since fractionation got worse.

MAXIMIZING BOTTOM PUMPAROUND		
	STEP 1	STEP 3
HEAT LOAD BOTTOM P.A. (GJ/h)	75.2	81.4
HEAT LOAD TOP P.A (GJ/h)	62.7	56.2
GAP HN/JF (°C)	9	11
GAP JF/LD (°C)	-23	-26
GAP LD/HD (°C)	-24	-27
FLASH POINT JF (°C)	49	48
E.B.P. ASTM D-86 JF (°C)	261	266

GAP HN / JF: ASTM D-86 temperature difference between JF 5% vol. and HN 95% vol. distillation points. Consider similar definitions for gaps between other products.

Step 4- Increase top pumparound to increase internal reflux and improve fractionation.

MAXIMIZING TOP PUMPAROUND		
	STEP 1 (adjusted *)	STEP 4
HEAT LOAD BOTTOM P.A. (GJ/h)	75.1	62.8
HEAT LOAD TOP P.A (GJ/h)	62.5	75.2
FLASH POINT JF (°C)	44	44
E.B.P. ASTM D-86 JF (°C)	244	239

Step 1 (adjusted *): Reduced crude tower top temperature by 5°C to reduce JF flash point and decreased 200 m^3/d JF rate to reduce JF EBP, since Step 1 was too far from JF specs.

Step 5- Increase top reflux to increase fractionation in the top pump around trays.

MAXIMIZING TOP REFLUX		
	STEP 4	STEP 5
TOP REFLUX RATE (m3/d)	3140	4098
HEAT LOAD TOP P.A (GJ/h)	75.2	58.7
GAP HN/JF (°C)	13	17
GAP JF/LD (°C)	-8	-4
GAP LD/HD (°C)	-10	-8
FLASH POINT JF (°C)	44	44
E.B.P. ASTM D-86 JF (°C)	239	240

Step 9- Drop some light diesel into heavy diesel to increase internal reflux in this region and test alleged positive influence on jet fuel quality, as reported by Operations.

KNOCKING DOWN LIGHT DIESEL		
	STEP 4	STEP 9
RATIO (LD RATE)/(HD RATE)	1 : 1	1 : 2
GAP HN/JF (°C)	13	16
GAP JF/LD (°C)	-8	-2
GAP LD/HD (°C)	-10	5
FLASH POINT JF (°C)	44	43
E.B.P. ASTM D-86 JF (°C)	239	233

Step 8- Inject stripping steam in the light diesel sidestripper to transfer light ends from this stream to the jet fuel stream and therefore knock down the heavy ends from jet fuel endpoint. Normally, Replan doesn't operate this sidestripper.

LIGHT DIESEL STRIPPER OPERATING		
	STEP 9	STEP 8
STEAM INJECTION (VALVE OPENING, %)	0	20
GAP HN/JF (°C)	16	19
GAP JF/LD (°C)	-2	-2
GAP LD/HD (°C)	5	9
FLASH POINT JF (°C)	43	42
E.B.P. ASTM D-86 JF (°C)	233	226

Step 10- Final optimization step. Incorporate any previous improvements obtained in earlier steps. In step 10, the jet fuel draw was increased, to maximize its yield. Recall jet fuel specs were 38 °C flash point and 225 °C end boiling point.

OPTIMIZATION		
	STEP 8	STEP 10
CRUDE TOWER TOP PRESSURE (kgf/cm2)	0.20	0.50
STEAM TO JET FUEL STRIPPER (t/h)	0.5	2.0
STEAM TO LD STRIPPER (valve opening, %)	20	50
JET FUEL RATE (m3/d)	1,600	2,000
GAP HN/JF (°C)	19	18
GAP JF/LD (°C)	-2	5
GAP LD/HD (°C)	9	1
FLASH POINT JF (°C)	42	39
E.B.P. ASTM D-86 JF (°C)	226	226

Steps not taken (2, 6 and 7) had been devised for the event of jet fuel presenting a low flash point, which did not happen. In step 2, more light naphtha was to be drawn from the pre-flash tower; in 6, reboiler heat to the jet fuel sidestripper would have been increased and finally in 7, the steam injection to this tower would have been increased.

The final results of the Unit Test were incorporated into the operational procedures of U-200A. This pushed the jet fuel yield to around 70% jet fuel yield obtained with trays, in following jet fuel runs.

4- Simulations

To evaluate the unit and to test the previously listed possible causes of column malfunction, the following simulations were run:

- Simulation of each step of the Unit Test, seeking the number of theoretical stages in each bed and the internal reflux rates in the column.
- Simulation of U-200 (the other Unit) in jet fuel run mode to assess the number of theoretical stages in each region of its trayed tower for comparison to the purchased stages of the packed beds of U-200A.
- Variations to the previous simulations to estimate the influence of the number of theoretical stages on the jet fuel yield.

Data on the number of theoretical stages employed in other projects with packing as well as best practices from knowledgeable authors were also gathered. Kister recommends 4-5 theoretical stages between heavy naphtha and jet fuel and the same between light diesel and jet fuel (Distillation Operation, 1989).

REGION	REAL TRAYS	PURCHASED THEORETICAL STAGES				
	1997 U-200	1997 U-200A	1999 U-200	2003 U-200A	1997 U-32
TOP P.A.	4	4 trays	4 trays	4 trays	1
HN x JF	7	3	4	4	5
JF x LD	10	4	10	11	6
BOTTOM P.A.	2	2 trays	2 trays	2	1
LD x HD	4	4	3	3	4
HD x AGO	-	5	4	-	1 (PA)
AGO x Overflash	4	3	3	4	4

Table 1

From table 1, we comment:

- Until 1997 U-200 and U-200A were operating fine in regard to jet fuel, with such gaps as 20°C between jet fuel and light diesel. Maybe because of that, we paid no attention to the efficiency of those trays, assuming the conventional 50%.
- The 1997 U-200A project had to squeeze an extra side draw (AGO) in the same column height, so there was no room for tall beds. Thus we reduced the top beds a little bit more. This project would be reproduced for the 1999 U-200 revamp project.
- After the jet fuel problem started, following the 1997 U-200A revamp, our "task force" studied column efficiency and found very surprising numbers for the trays in the jet fuel x light diesel region. See table 2.
- We therefore recommended a revision for the 1999 U-200 revamp as in table 1. One of the main points was to increase the number of theoretical stages between jet fuel and light diesel. For Replan, jet fuel is a most important product.
- In 1999, U-200 started up with this and several other modifications recommended by our group. Despite all preventive steps taken, U-200 performed just as bad as unit U-200A in regard to jet fuel yield. Of course, more studies were required, and we are confident that we now know what went wrong. But we will only tell you after we confirm it in the next turnaround. Clue: liquid maldistribution!
- 2003 U-200A started up in July and is performing above design. What it took to solve it will be detailed in the next sections.
- 1997 U-32 belongs to another Petrobras refinery. It was a grassroots distillation unit, and ever since its start-up, presented the same jet fuel problem as U-200A and U-200. It returned from it first turnaround in January 2004, and the jet fuel yield is now equivalent to a trayed column from the same refinery... Good news!

REGION	SIMULATION THEORETICAL STAGES				
	1997 U-200	1997 U-200A		1999 U-200	2003 U-200A
		STEP 4	STEP 10		
HN x JF	3 to 4	1	2	4	Over 4
JF x LD	8 to 9	3	4	4	Over 11

Table 2

From table 2, we comment:

- 1997 U-200 – Top trays achieving 40 – 60% efficiency, as expected, whereas jet fuel trays boosting 80 – 90%, wide over expectations. It means we had a big loss in stages when moving from trays to packing.
- 1997 U-200A - Conditions proposed in step 10 led to column overpressure, so unit typically operated somewhere between steps 4 and 10. Either way, both beds were underperforming, by comparison to purchased stages (table 1).
- 1999 U-200 – The 10 purchased theoretical stages were delivering just 4. What happened?
- 2003 U-200A – Now we got more than we bought…Great!

Concerning unsuccessful revamps from trays to packing in distillation towers, two external references were found. One was in Russia, very similar to the U-200A case. The root cause was liquid maldistribution. It was fixed changing liquid distributors. The other one happened in a Petrobras grassroots C_3 splitter, which could not specify product purities. Changing the pan distributors to gravitational ones made the tower perform above design.

We studied the influence of theoretical stages on the jet fuel yield. This was done by simulation fixing two specs of jet fuel; the 10% volume distillation (which correlates well to flash point) and the final boiling point. We found a big influence from jet fuel x light diesel bed stages, but no significant influence from heavy naphtha x jet fuel stages.

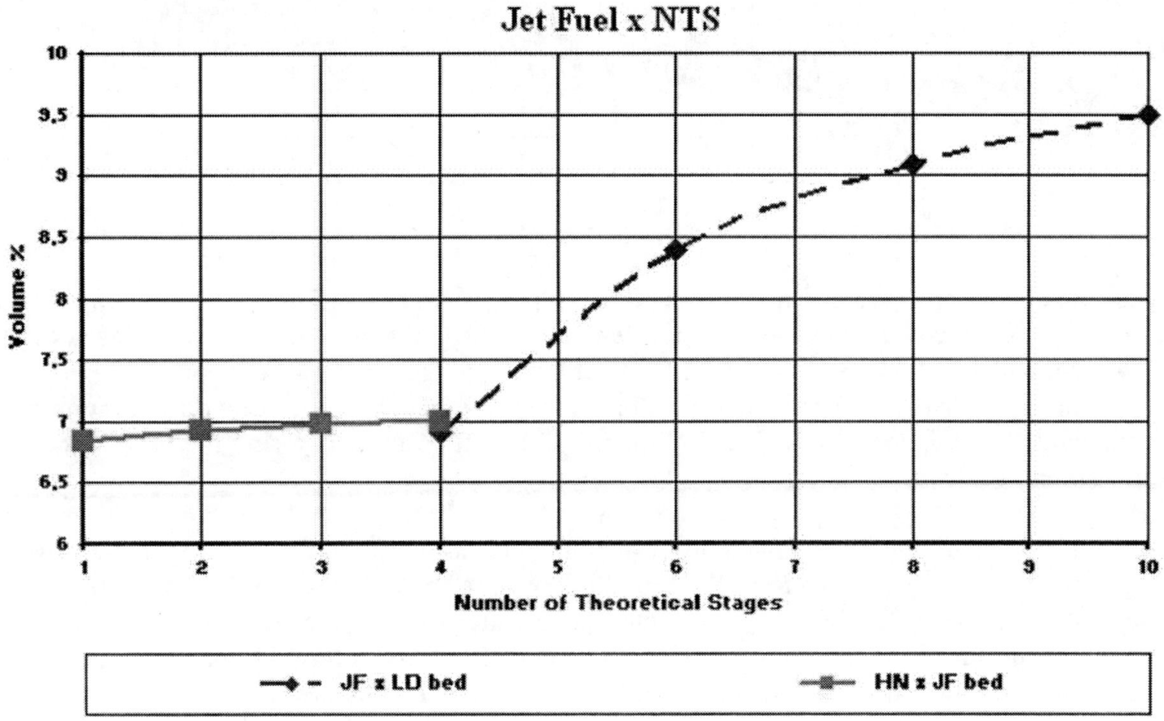

5- Internals Evaluation

We did a hydraulic check on collectors, distributors and packed beds for Design and Unit Test conditions. The criteria were: residence time, nozzle velocities, down pipe self-venting capacities, distributor levels, packing flood and flux parameters.

In this column we included level gauges in all liquid distributors. Based on them we could infer internal reflux rates and compare them to simulated and design rates. None of the Unit Test steps exceeded the design rates, which means the internals should not present capacity problems at all. Inferred internal rates were about 20% higher than simulated for the two top beds and 10% lower for the third bed. We considered this an excellent agreement.

The distributor above JF x LD bed showed strong level variation (pulsation pattern), even at 50% unit charge rate, spanning about 20%, cycling every 3 to 5 minutes. The chart below reports distributor levels JF x LD (LI20527) and LD x HD (LI20528), for comparison, during the Unit Test.

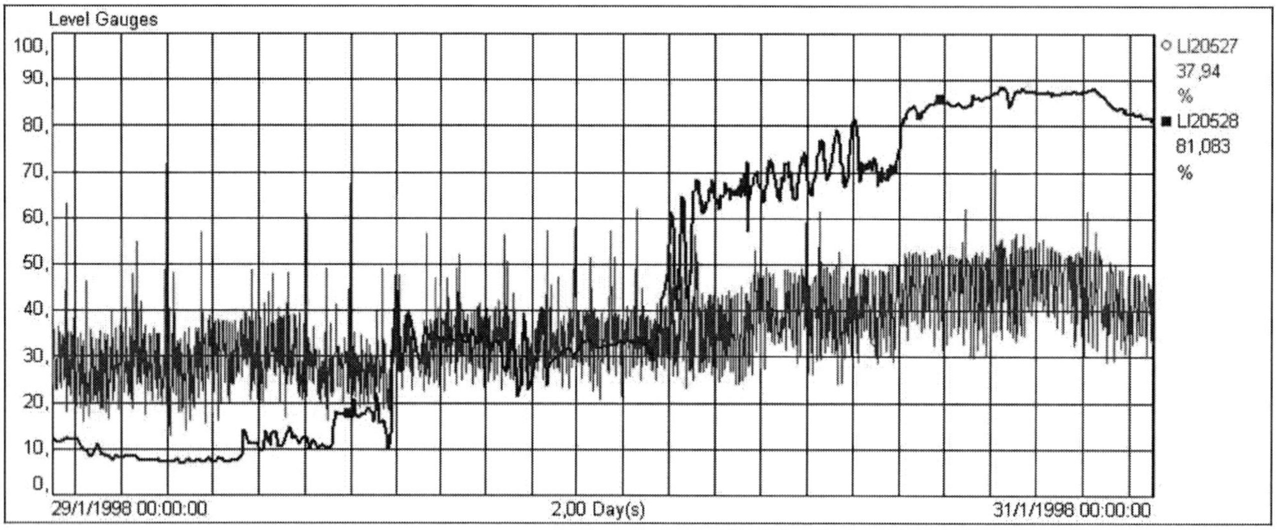

The hydraulic checks revealed that none of the down pipes were self-venting. Most downpipes had an entrance restriction which amplified the problem. Also, the residence time criterion was not met for the top two collectors (HN x JF and JF x LD), although we feel this is a very conservative guideline. The geometry of the three top collectors is troublesome, with a narrow do-nut mounted against the wall. The Vendor used the same dimensions for all do-nuts, ignoring the big difference in flow rates among sections.

Level pulsation in distributor JF x LD could be attributed to this combination of facts. Physically, the downpipes are non self-venting, causing the liquid to "pulsate" up and down in the very small do-nut collector system that feeds the JF x LD distributor. The collector may overflow, throwing liquid right onto the wall, the worst situation for any packed bed. The thrown liquid bypasses the entire bed, ending up at the light diesel draw off collector. The liquid that reaches the JF x LD distributor comes in gushes, causing the pulsation in this distributor, which may also overflow.

Hydraulic checks quickly and effectively link simulations to field data (the Unit Test). They give clues as to the limiting internals and indicate where to focus gamma scan surveys. Hydraulic checks are normally presented in literature. (Henry Kister, Norman Liebermann, Andrew Sloley, Scott Golden, FRI Fractionation Research Institute, etc)

6- Gamma scan

The gamma scan surveys intended to look for liquid distribution problems in the packed beds and liquid overflow in collectors and distributors. The scans were conducted in two different conditions, maximum flow rate in diesel run mode (28,000 m^3/d), and jet fuel run mode at a lower feed rate (24,000 m^3/d). The last condition requires higher reflux rates to get better fractionation.

Findings:
The scan results were very useful in helping confirm previous conclusions. Most regions could be analyzed in good detail. A few regions had some interference which reduced the opportunity for a reasonable understanding.

- The do-nut collector below top bed (HN x JF), which feeds distributor JF x LD, confirmed liquid overflow in both scan conditions. The scan technicians suspected of level changes between scan lines (or would it be the pulsing behavior described above?) Anyway, this means liquid going to the walls and bypassing the JF x LD bed.
- The JF x LD distributor was probably overflowing on both conditions. As explained earlier, we suppose this originates from the pulsating liquid coming from above.
- On both conditions, the JF x LD bed exhibited a density gradient, starting with a big accumulation of liquid on its top 0.6 m, which progressively redistributed itself as it traveled downwards. The 0.6 m represents 30% of total bed height which are useless for fractionation, because it is flooded. Our team discussed the possibility of this being a liquid ring formed by the overflow of the JF x LD distributor. This seems to be the case, as later indicated by a Cat Scan™ performed 0.6 m below the top of this bed. The JF x LD bed had liquid surging though it during the Cat Scan™. Moreover, the radiation intensity passing through the column was oscillating up and down, just as the liquid level on the distributor. The density profile showed the extra liquid from the surges was going two ways: down the walls and through a central circle with radius about a sixth of the tower diameter. Presumably, this central circle is linked to overflow from the JF x LD distributor, which has a main central hexagonal channel feeding secondary arms.
- The do-nut collector below JF x LD bed overflowed in the second condition. No clear vapor line was seen above the do-nut, meaning heavy compounds from light diesel were being entrained to the bottom of the JF x LD bed, one extra reason for bad fractionation.

The other sections of the crude tower were found normally positioned and with no particular signs. This also applies to the jet fuel sidestripper, whose trays were operating without entrainment nor weeping. Besides, the sidestripper liquid level was found normal, about 2 m below the steam injection point.

7- Modifications to U-200

By this time, the project for other unit, U-200, had datasheets out for bidding which followed the same concepts from U-200A, so something had to be done to avoid duplicating the problem. Due to the tight schedule, the "task force" analyzed alternatives and decided to keep the original configuration (trays in pumparound and packing in fractionation regions) for U-200, making the required adjustments on datasheets and recommending some improvements.
All Petrobras supplies are done by bidding, so for U-200 the column package was delivered by a different company than for U-200A.

Data Sheet revisions:

- Do not accept do-nut shaped collectors; the recommended configuration is chimney trays.
- Require self-venting downpipes, if residence times for internal reflux rates are below 30 seconds.
- Since jet fuel is very valuable for Replan, we oversized the second bed (JF x LD) advising 10 theoretical stages, and added one stage to the first bed (HN x JF).

Assembly recommendations:

- Check all dimensions of internals supplied;
- Extensive leveling during assembly;
- Collector leak tests and distributor arms leveling with water;
- Water test, covering maximum and minimum operational rates, for collectors and distributors.

Assembly Results:

- Internals and installation checks all done;
- Due to water pumping limitations, the maximum operational rates could not be achieved in some regions;
- Detected design problems in two collectors, JF x LD and LD x HD. The collector chimneys were transverse to main flux. Moreover, big triangular chimney reinforcements severely hindered flow, causing chimney overflow.

Those reinforcements were inadequate and unacceptable. Some other solution should have been sought in advance (thicker plates; under tray beams, etc.). But when project teams don't look at it beforehand, there will be no alternatives left, once the tower is open and Maintenance is on the final countdown to shut the tower.
Field solution:

- JF x LD collector chimneys height was raised 60 mm. This was the sole correction allowed by the turnaround schedule. Operations restarted the following days.

Results:

Since the 1999 revamp, U-200 began presenting low jet fuel yields. The flash point results were better than U-200A, but the endpoint contamination was just as bad. Light Diesel contained at least 30% jet fuel fractions.

After tests, simulations, evaluations, gamma and CatScans™, the conclusions pointed towards liquid maldistribution again, in beds JF x LD and LD x HD, exactly the two with collector design problems...

Petrobras has devised two possible solutions for U-200, but we have to wait for the next turnaround (2005) to implement them.

8- 2003 U-200A revamp

After such a disappointing performance from U-200 regarding jet fuel, we had to think it over. As we studied the U-200 case, we realized the whole revamp process was a chain, and the failure of any one link ruins the final result.

We started looking at the upcoming revamp of U-200A, originally planned for 2002, seeking another upgrade in capacity (30.000 m^3/d) and a still heavier slate. Lacking any good reasons to believe that this time it would work, the first thing we did was go back to the drawing board. We then came up with a new outline for the design, which eventually became the backbone for the actual revamp (postponed by other reasons to 2003). The new approach was:

- New vacuum furnace - To handle the heavier slate and the diesel spec trend, much lighter;
- Eliminate the AGO drawoff - At 30.000 m^3/d and no changes to atmospheric furnaces, AGO draw would overload them. Besides, AGO was already not in use and the new vacuum furnace could be designed to handle it;
- Move HD draw to former AGO drawoff;
- Move LD draw to former HD draw; this means the tower would have two consecutive JF x LD beds;
- Stick to a high number of theoretical stages for JF x LD fractionation, requiring at least 8 stages;
- Together, these 4 last actions guarantee increased wetting in the new LD x HD and the second JF x LD beds;
- The bottom pumparound region was changed from trays to packing, using a spray distributor.

All these outlines, together with pertinent data, were sent to the original U-200A packing supplier, requesting a diagnosis of the current situation plus their proposed solutions for the 2003 column revamp. After several technical meetings, Petrobras and the supplier agreed upon a final concept and datasheets by September 2001. The only changes from initial Petrobras outlines were:

- Permit joining the two consecutive JF x LD beds into one bed, as there was no product to be drawn between them;
- Advise a gravitational distributor for top pumparound bed, to achieve an appropriate distribution for the internal reflux, which accounted for 50% of total flow.

One shortcoming, the supplier did not explore any new possibilities for the future configuration, limiting himself to those Petrobras outlines. The supplier did admit overflow of the do-nut collectors as the root cause for current column malfunction, but rejected it being undersized, rather bringing up other possible culprits: foaming, plugging, damage at start-up, none of which proved true after opening the column.

Petrobras bid the crude tower revamp along 2002 and this time a third company was awarded the contract. Their project followed the datasheet outlines with the following features:

- The two consecutive JF x LD beds were joined, delivering 11 theoretical stages;
- The company included a proprietary feed distribution device (vapor horn).

Comparing the 2003 solution with the former ones, there are some conceptual differences we believe may be adding some benefits:

- The downpipes were changed to down boxes, increasing area and therefore avoiding self-venting problems;
- The collector solution employs a larger number of smaller chimneys, reducing their total area from 23% to 13% of tower cross sectional area and therefore providing better vapor distribution;

The assembly of internals went well. Although this time we had not scheduled wetting and water tests as in U-200, fact is after start-up the jet fuel yield went up to the theoretical maximum, similar to the former trays performance.

Table 3 compares U-200A field data before and after the 2003 revamp.

		1997 REVAMP		2003 REVAMP	
		Step 4	Step 10	11/22/2003	2/12/2004
Column top pressure	kgf/cm2	0.1	0.5	0.4	0.3
Column top temperature	℃	106	106	103	108
Jet Fuel draw temperature	℃	157	161	170	174
Naphta yield (LN+HN)	%vol	14.5	24.6	10.0	25.9
Jet Fuel yield	%vol	6.7	7.9	10.1	10.2
Diesel Yield (LD+HD)	%vol	24.7	24.6	31.9	29.4
Jet Fuel end point	℃	239	226	224	221
Jet Fuel flash point	℃	44	38.5	45	42
GAP Naphta x Jet Fuel	℃	12.7	18.0	33.6	29.3
GAP Jet Fuel x Light Diesel	℃	-8.3	5.2	30.3	22.4
% LD distilled at 225ºC	%vol	12	18	<2	5
				Heavy slate	Regular slate

Table 3

9- Lessons Learned

Three great lessons were learned from the recurring jet fuel problem.

Number 1: **Start your revamp with correct outlines**.
When revamping any unit, start by assessing the actual performance of the equipment out there in the field. The best way to do this is by running a unit test and then simulating it to find out actual efficiencies, capacities, constraints, fouling factors, etc. Seems obvious, no? But if we had done this for the 1997 U-200A revamp, we would have certainly realized that the existing trays delivered almost 90% efficiency, not the conventional 50% used for design. This would have prevented Petrobras from designing the JF x LD section with 4 theoretical stages instead of 8 or 9. Correct revamp outlines are essential for ultimate success.

Number 2: **Great process solutions can fail if mechanical design is not in line with process.**
We saw this in the 1999 U-200 case: an overlooked detail such as those chimney reinforcements created an unexpected barrier to liquid flow. Final result: the 10 stages that were purchased for the JF x LD bed were reduced to 4. This was just one example among numerous other that can ruin performance if they don't work together with process. There is a second reading from the Lesson 2: packing can be fabulous, but it is vulnerable to a lot more factors than trays. Trays naturally redistribute liquid and vapor at each one of them, so random errors do not add up to a systematic error. In this sense, trays are more robust than packing. Therefore, we restate Lesson number 2 as:
Great PACKING solutions can fail if mechanical design is not in line with process.

Number 3: **Liquid distribution was the key issue in all cases presented in this paper.**
In the 1997 U-200A revamp, liquid distribution was part of the problem, whereas in the 1999 U-200 revamp it was the real singular problem. In other units mentioned in this paper (Russia, Petrobras C_3 splitter and U-32), liquid distribution was the villain. Finally, we point out that in U-200A and U-200, the specific cause of liquid maldistribution lied not on the distributors themselves, but on collectors overflow.

10- Acknowledgements

1997 "Task Force":
Antonio Ricardo Lanfredi
Eduardo Veloso
Gilberto Aoki
Gláucia Torres
Gunter Schnaibel
Maria Elizabeth Marsiglia

Special thanks to Henry Kister for his assistance during troubleshooting and to Luis Claudio Moreira Paschoal from CENPES for his work on the 2003 U-200A basic design.

Last but not least, we thank the help from the Replan Team.

Attachment 1 – Original columns sketch (U-200A and U-200)

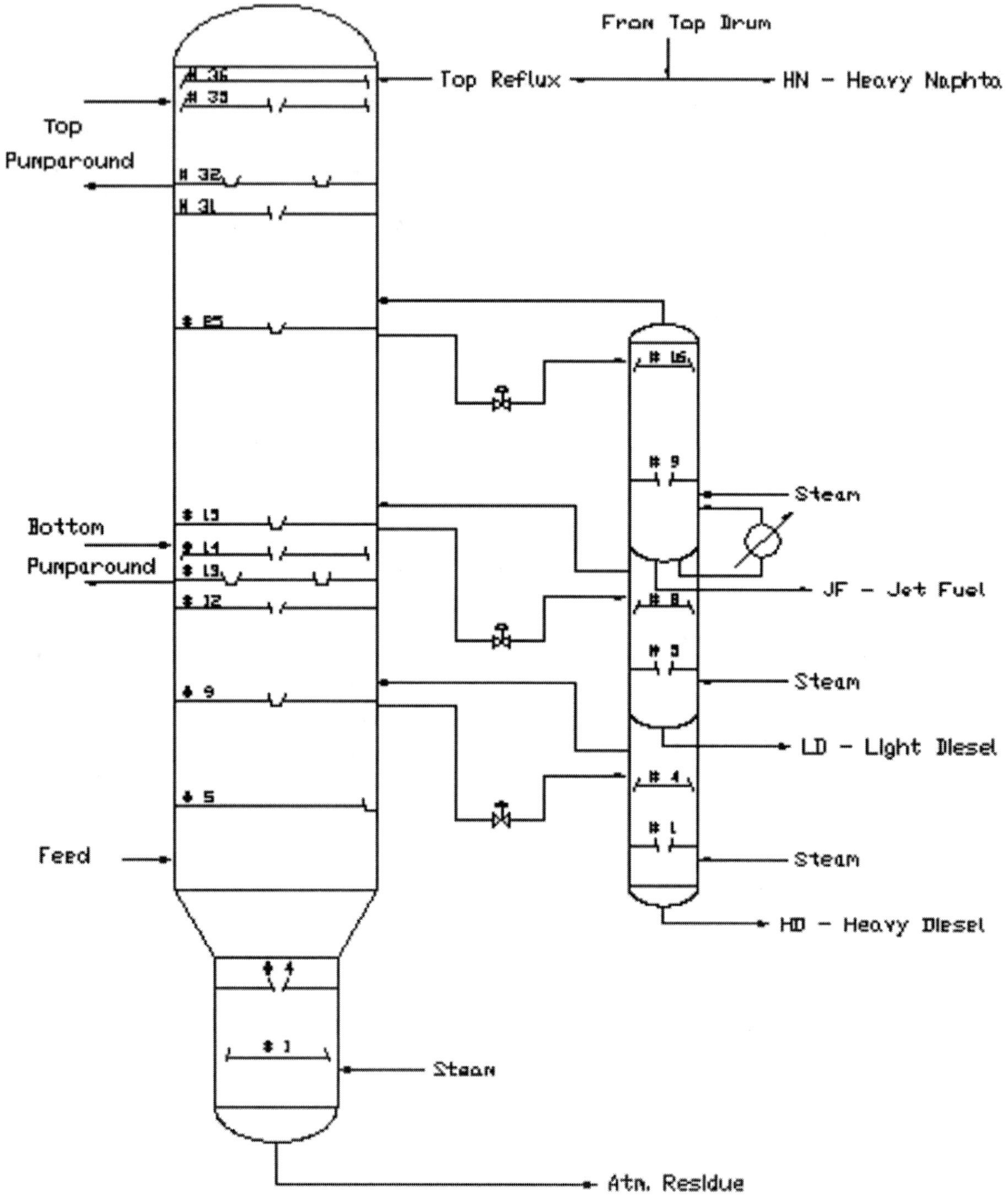

Attachment 2 – U-200A Crude Column after 1997 revamp

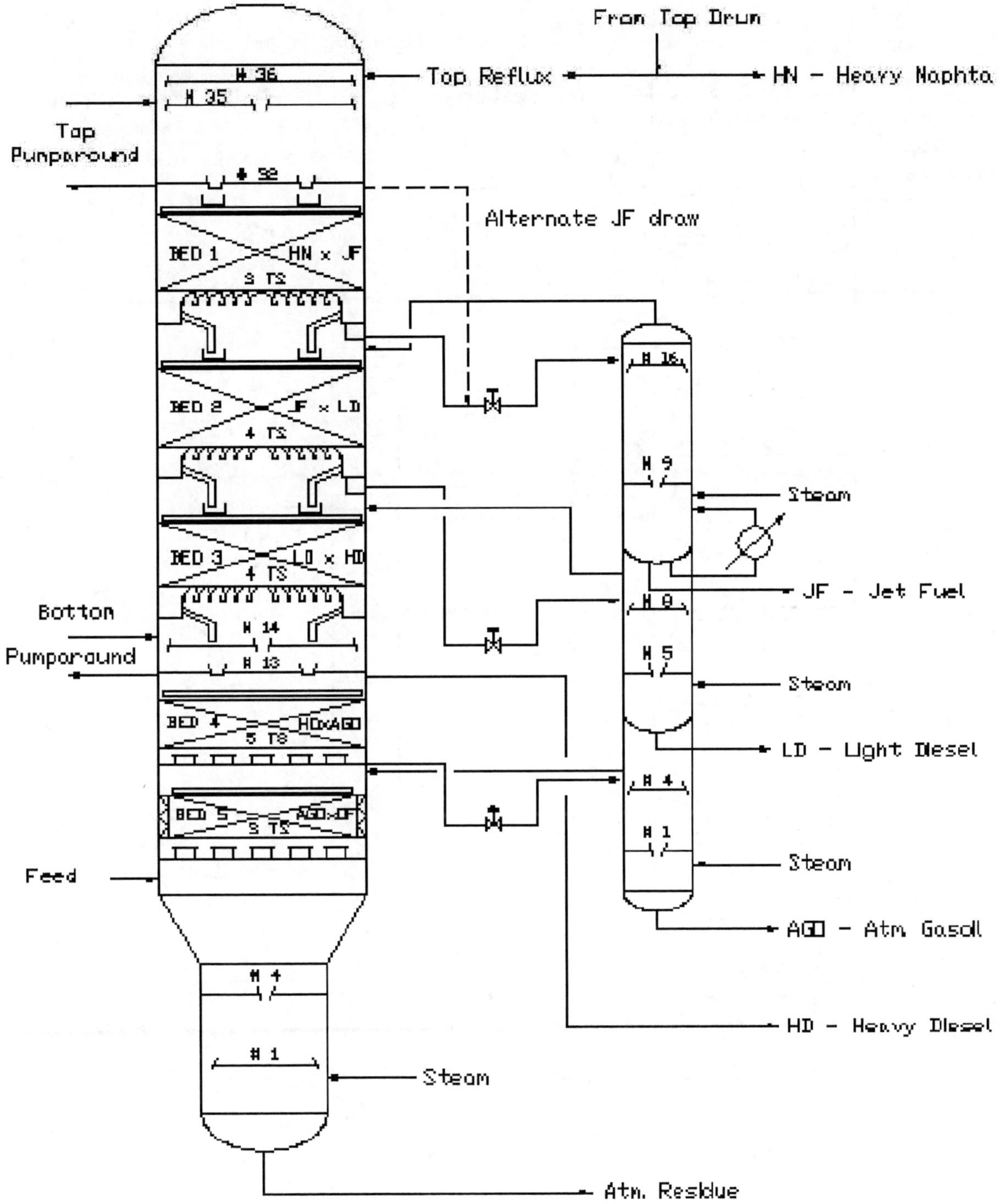

Attachment 3 – U-200 Crude Column after 1999 revamp

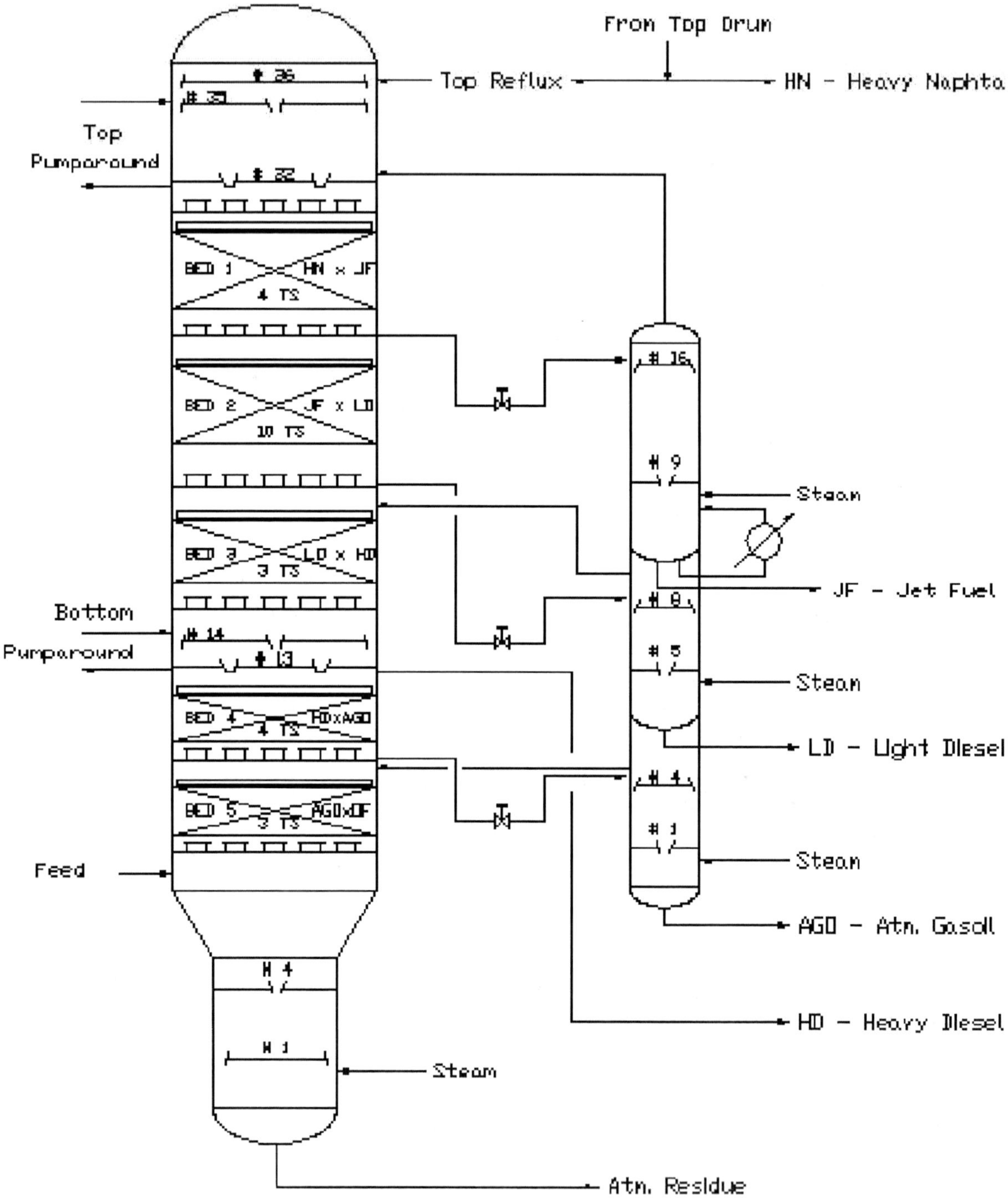

Attachment 4 – U-200A Crude Column after 2003 revamp

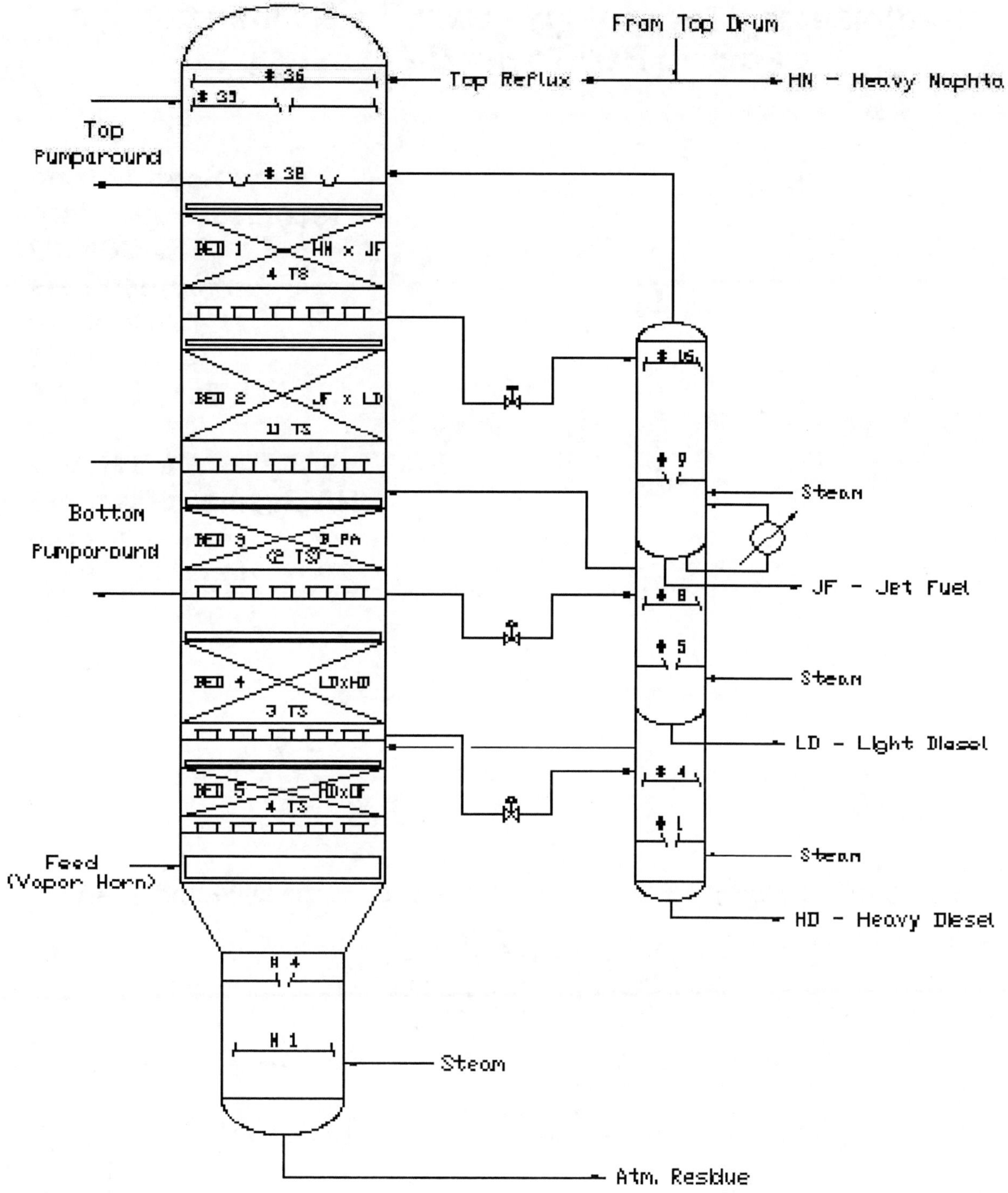

Radioisotope Technology - Benefits & Limitations in Packed Bed Tower Diagnostics

By: Scott Vidrine
Technical Supervisor
TRACERCO
1100 Hercules
Houston
Texas 77058
Tel: 281 488 0039

Paul Hewitt
NA Business Manager
Tracerco
Houston, Texas

Presented in the Ethylene Producers Conference of the 2004 Spring AIChE Meeting
New Orleans, LA.
April 21-April 23, 2003

Radioisotope Technology - Benefits & Limitations in Packed Bed Tower Diagnostics

Introduction

The use of radioisotope technology in Chemical, Petrochemical and Refinery Process Plant mass transfer vessels has continuously been developed over the past 50 years to assist engineers in troubleshooting process problems, optimize production and plan shutdown schedules to minimize plant down time. Many advances in data acquisition and processing ability now allow critical information to be provided quickly to ensure the most efficient use of plant equipment.

With the advancement of data gathering and handling it is important to note that fundamental physics behind each and every measurement involving radioisotopes has never changed. As with any technology there are many advantages in its use. However there are also limitations that a process engineer must fully understand so that a possible problem is not overlooked or simply discounted based upon results generated.

Mass transfer design can typically be split into two main categories; trayed and packed beds. Packed beds are becoming increasingly popular. In the case of a trayed tower, the open area is typically less than 10% whereas the open area in a packed bed is greater than 50%. Additionally, packing tends to have a lower liquid holdup than trays. These characteristics combine to result in pressure drop for a theoretical stage being lower in packing than in trayed systems. This reduction in pressure drop can have a significant impact on tower and plant economics.

As with all situations, packed towers bring their own particular challenges. As an example, the same open area and reduced liquid holdup that benefits pressure drop may result in lower liquid and vapor velocities, and therefore, an increased susceptibility of packed beds to accumulate solids in systems prone to polymerization or precipitation. Another common issue surrounds mechanical integrity. Packed bed towers are more susceptible to damage as a result of pressure surges as compared to trayed vessels. Finally, a critical consideration in the effective operation of a packed tower is the mechanism for fluid distribution. Poor vapor or liquid distribution can result in a significant efficiency reduction. In fact, the necessity of uniform distribution increases with the number of theoretical plates in a bed. This results in tower internal designers limiting bed length and redistributing process materials at specific and regular intervals.

With the increased use of packed beds the application of radioisotope technology to troubleshoot problems has become popular. This paper, with the aid of case studies, describes the general principles of the radioisotope diagnostic technologies used in both trayed and packed bed towers that allow operating characteristics to be measured. It shows the benefits that can be gained by using the technologies as well as limitations and in some cases the necessary strategy of radioisotope technology integration to successfully determine the source of a particular mass transfer problem.

Outline of Radioisotope Technologies Used in Tower Diagnostics with Case Studies

Gamma Ray Transmission Scanning

The intensity of detection of a source of gamma emitting radiation is dependent upon several factors including the source type used, the distance between the source and detector and the density through which the source has to pass to reach the detector. If the source and detector are positioned at a fixed distance apart the detector response is proportional to the density it encounters in-between. This physical parameter can be used to determine the density profile through a vessel. Relating the density profile to expected materials within the vessel allows diagnostics to be performed on the vessel contents.

In the case of a tower, a source can be positioned on one side and a radiation detector positioned horizontally opposite, on the other side. In the case of a trayed tower the radiation beam is normally passed across the active area of the trays (avoiding the downcomers) to determine the liquid level on the tray and the quality of the liquid/vapor disengagement above the tray. The source and detector are then moved vertically downward and the intensity of the radiation transmitted through the tower is recorded at each measurement position.

As the source and detector move downwards through a normal vapor space within a tower, a transmission peak (high intensity) is observed, and when the scan line intersects liquid on a tray, a packed bed, or other dense material, an absorption peak (low intensity) is observed.

In practice the changes in scan response between regions of low intensity to high intensity, or vice versa, are not perfectly sharp because wide beam instead of narrow beam radiation is used. It is necessary to use wide beam radiation because the shielding required to produce narrow beam radiation would make the source container too heavy and unwieldy for practical use. Consequently, there is a contribution to the radiation field at any point due to radiation scattered indirectly from the internal structure of the tower toward the detector.

A graph of radiation transmission measurements as a calculated relative density versus tower elevation generates a density profile of the operating tower. Interpretation of the density profile can identify a variety of operational and mechanical problems, such as flooding, entrainment, foaming and damaged internals. The density profile is normalized for diameter changes and wall thickness variations.

All quoted density values are calculated relative to the highest observed radiation transmission, which is assigned a vapor density of zero pounds per cubic foot. The quoted vapor density represents the minimum observed density in the vapor disengagement zones, and indicates the quality of the liquid/vapor disengagement. The tray aerated liquid densities present the maximum observed density of the material on the tray deck, including any tray hardware and is an indication of the amount of aeration taking place on that tray. Likewise, the quoted aerated liquid heights include absorption due to aerated liquid on the tray, the tray itself, and the tray supports and indicate the liquid level on the tray as well as the amount of aerated liquid, foam, or froth above the liquid.

One-half the averaged tray maximum liquid density is used as the tray liquid/vapor interface point. That is, all density responses heavier than this point are considered essentially aerated liquid, and all density responses lighter than this point are essentially vapor

Trays with liquid density less than the indicated interface reference density are usually considered operationally dry. Dry trays can be due to operational conditions such as entrainment lifting liquid from the trays due to high vapor velocities, liquid flashing conditions due to high temperatures, low liquid reflux or feed rates, and liquid draw points.

Different liquid hold up levels may also be due to different tray designs such as sieve, valves, bubble caps and or downcomers in different sections of the tower. Once all of the operational and design considerations have been eliminated, marginal liquid levels may be attributed to tray mechanical problems.

Case Study – Gamma Transmission Scanning

PROBLEM

A major oil refiner was experiencing a high differential pressure in the main fractionator column limiting production to one half normal rates. The planned turnaround was six months away with major revamp work planned with the replacement of the stripper section to increase unit capacity. An unplanned turnaround would triple the planned cost. A tower scan was needed to identify the cause and location of the liquid restriction.

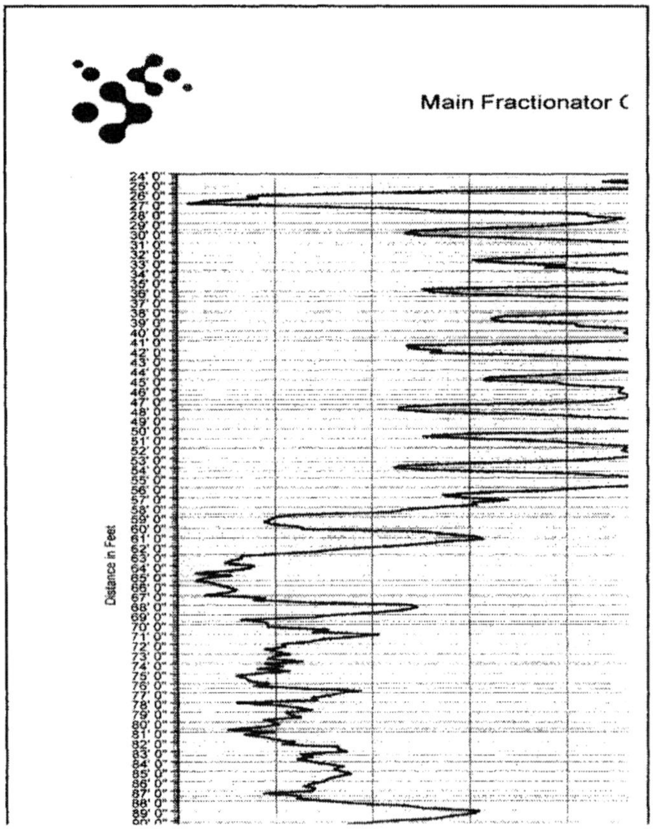

RESULTS

Within hours of job scope discussions with the refiner, a team of two engineers performed an initial tower scan and located the liquid restriction two trays below the bottom chimney tray (see Figure 1: Red Solid Line).

The location of the liquid restriction appears to begin in the downcomer of tray 8. It is obvious that the liquid restriction has caused a liquid stacked flood extending from tray 8 through tray 16. A liquid stacked flood can be described as a significant increase in the maximum liquid density with no liquid and vapor disengagement between trays.

Existing piping was identified as a potential pumparound to bypass the restricted trays and prevent a shutdown. Once the pumparound was put in operation, a second tower scan was needed to help determine the effects of the new pumparound on tower operations (Figure 2: Green Dashed Line).

Fig. 2

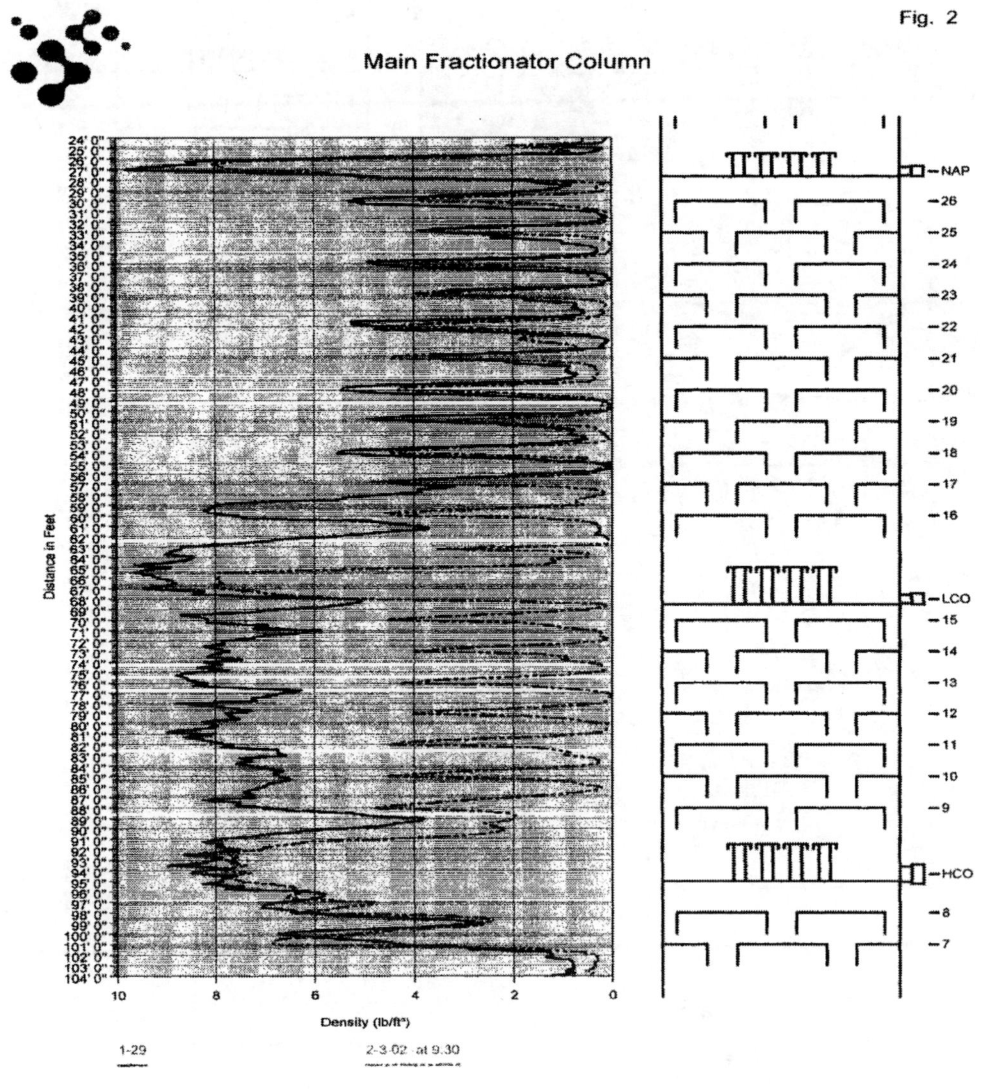

The flooding had been significantly reduced with an increase to two third previous rates. The chimney tray and tray 8 were still flooded indicating the 4 inch pumparound was still not enough. The pumparound was hydraulically limiting in only a 4 inch line available, so a second 4 inch line was added to increase the potential liquid volume. Once in place and operating, the unit was able return to full operating rates. A final tower scan was performed to confirm the tray integrity (Figure 3: Blue Dashed Line).

Fig. 3

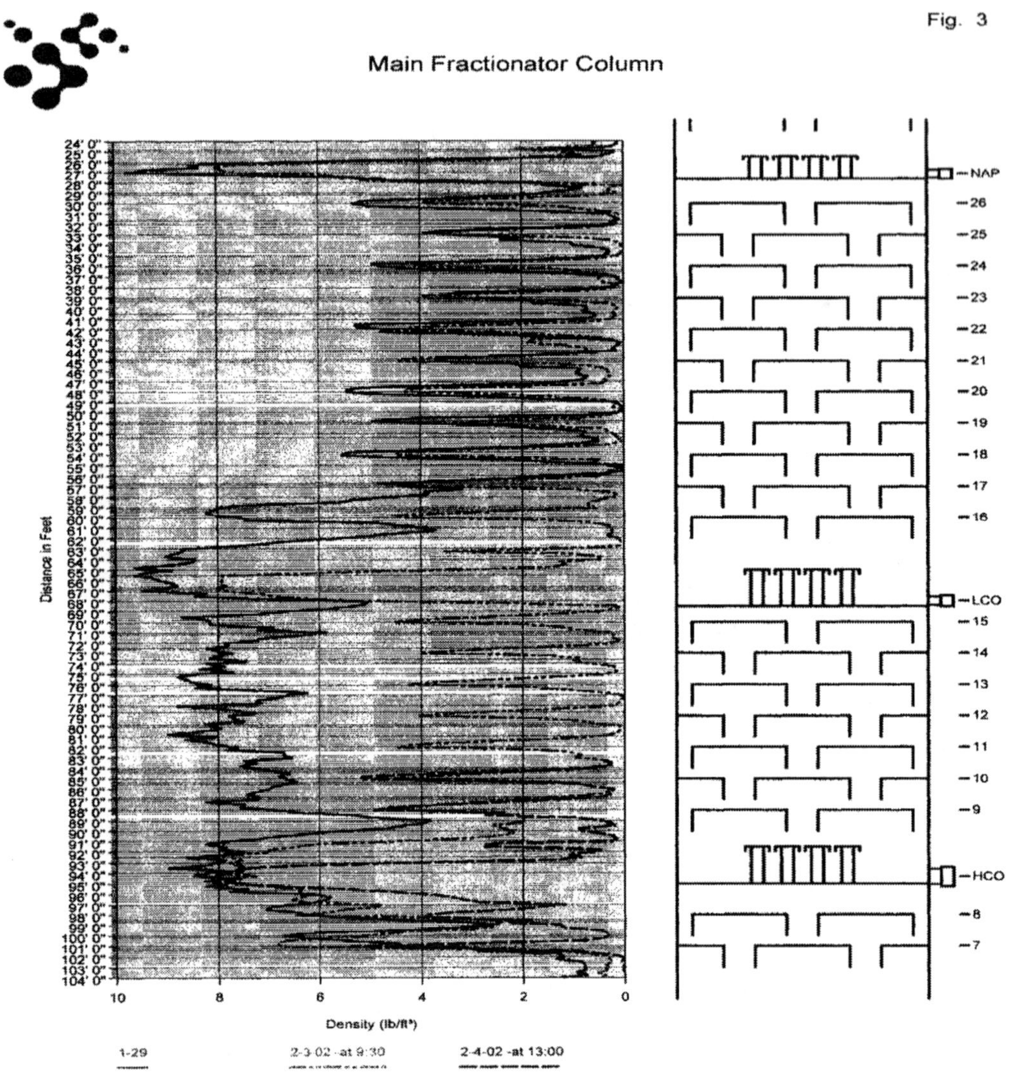

Radiotracer Residence Time & Distribution Measurement

Injecting a compatible radiotracer into an appropriate inlet upstream of a vessel and monitoring its passage through the vessel allows a residence time and fluid distribution to be measured. Sensitive radiation detectors are placed at strategic elevations / locations on the vessel. The detectors are relatively small and easy to mount at each position. Each one is connected by a cable to a central data logging device that records radiotracer concentration versus time information.

When the radiotracer passes each detector a response is registered and recorded. Prior to the test each detector is assessed and its response normalized such that each detector responds identically to a given unit of radiotracer. Analysis of each detector response allows significant information to be obtained including time of flight, pulse duration, Inverse Peclet Number, Stirred Tank Equivalent, individual detector response area and the individual percentage from each detector response area relative to the total of response areas for each distribution ring of detectors. This in turn allows maldistribution to be determined.

In the case of a packed bed tower detectors are usually placed towards each of the individual packed beds as well as the inlet and outlet pipe work. In the case of liquid flow the upper detector ring allows determination of any mal distribution due to the liquid distributor. The lower ring of detectors allows the detection of mal distribution from within the packed bed. In the case of gas this is reversed with the lower ring detecting any distributor mal distribution issues and the upper ring any packed bed distribution effects. The outlet and inlet detectors allow residence time to be measured and compared to calculated values. They also allow the presence of channeling to be further verified by the mean residence time determination and the presence of concentration versus time "fingering" effects that show flow channels within each bed.

Tests are routinely carried out with packed bed towers using both liquid as well as gaseous tracer materials.

Case study – Maldistribution within a Packed Bed Tower

PROBLEM

A major oil refiner was experiencing poor performance from a newly installed packed bed in the base of a Main Fractionator Tower. A radioisotope distribution study was agreed to offer the best information to identify the cause of the poor performance.

RESULTS

A total of 14 detectors were mounted on the tower. Two detectors were used to monitor the inlets and outlets of the tower for timing. The remaining twelve detectors were mounted in two rings of six. The detectors were equally spaced every 60 degrees around the circumference of the tower. The top ring of detectors was mounted 3'4" below the slurry pump around or 1 ft into the packed bed. The bottom ring of detectors was mounted 7'0" above the vapor inlet or 1ft below the packed bed.

The detector orientations are described in Figure 4. There locations started at 0 degrees (above the vapor inlet) and were positioned every 60 degrees. For simplicity, the detector locations will be referred to as 0, 60, 120, 180, 240, and 300.

Where detectors were used to monitor flow distribution, a comparison of the detector response curve areas was obtained, with the total observed at the elevation by all detectors representing 100 percent. The expected distribution percentage per quadrant was **16.6** percent for equal distribution.

Figure 4

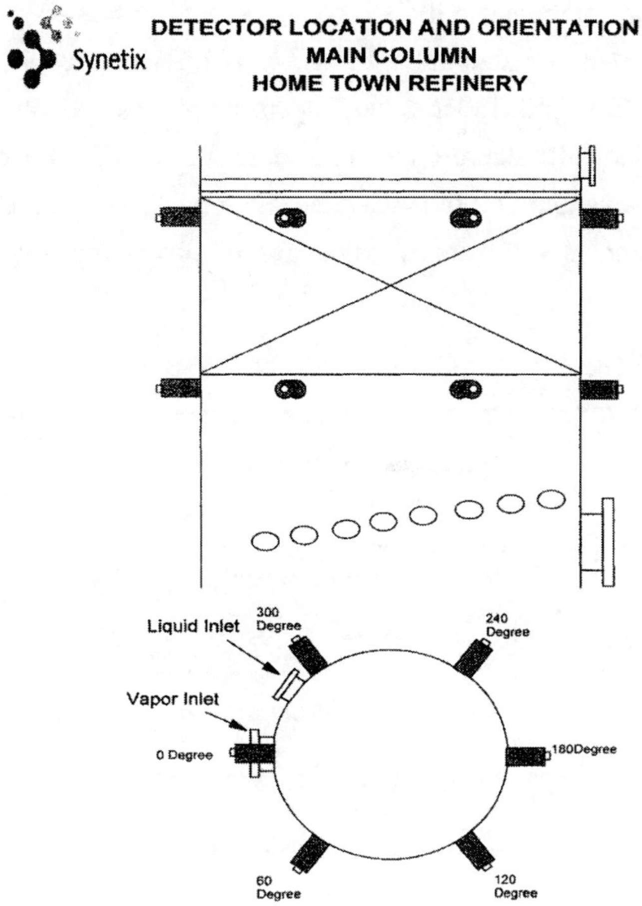

Vapor Distribution

The analysis of the vapor distribution in the bottom ring of detectors showed preferential

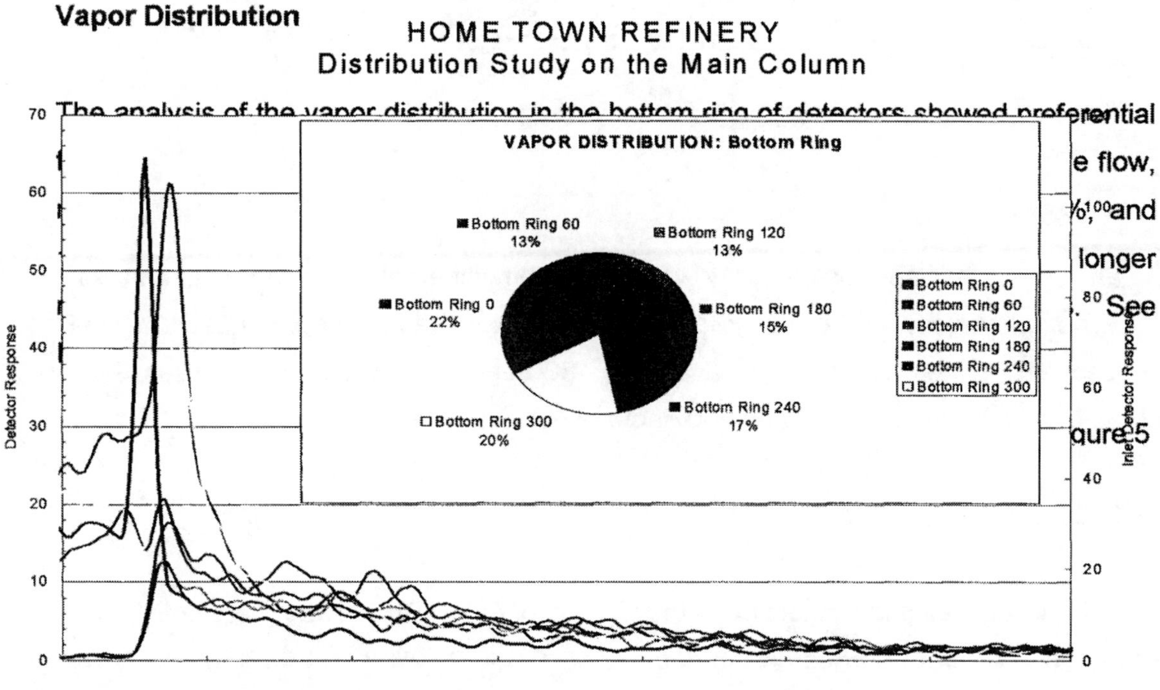

The analysis of the vapor distribution in the top ring of detectors showed preferential flow to the 0, 240 and 300 degree quadrant with 18.3%, 22.3% and 23.5% of the flow, respectively. The 60, 120, and 180 degree quadrants showed 12.6%, 12.5%, and 10.7% respectively. The 240 degree quadrant detector showed noticeably longer response tail and noticeably later in time indicating more backmixing as compared to 0 and 300 degree detector. See Figure 6 below for graphical representation of the response curves.

Figure 6

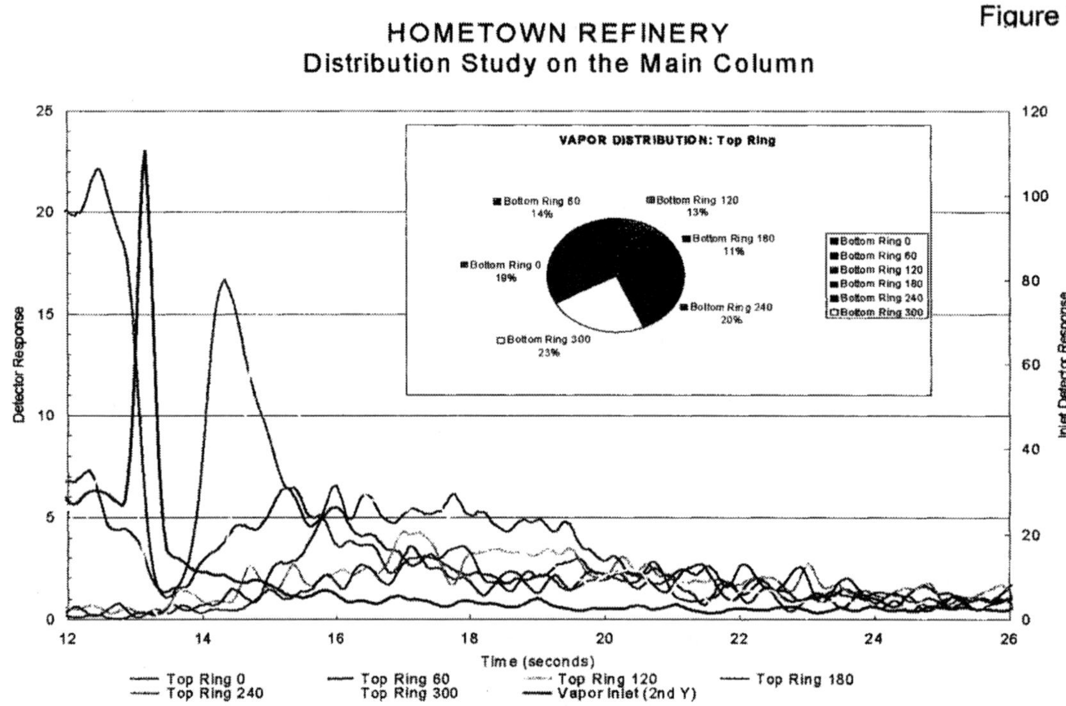

The analysis of the vapor rise velocity was significantly different within the quadrants as well. The rise velocity for the 0 degree quadrant was 4.4ft/sec, 60 degree quadrant was 4.2ft/sec, 120 degree quadrant was 2.9ft/sec, 180 degree quadrant was 2.6ft/sec, 240 degree quadrant was 3.2ft/sec, and 300 degree quadrant was 7.2ft/sec.

Liquid Distribution

The analysis of the liquid distribution in the top ring of detectors showed preferential flow to the 120, 240 and 300 degree quadrant with 20.5%, 22.2% and 20.7% of the flow respectively. The 0, 60, and 180 degree quadrants showed 8.7%, 16.7%, and 11.2% respectively. See Figure 7 below for graphical representation of the response curves

HOME TOWN REFINERY | Figure 7

Distribution Study on the Main Column

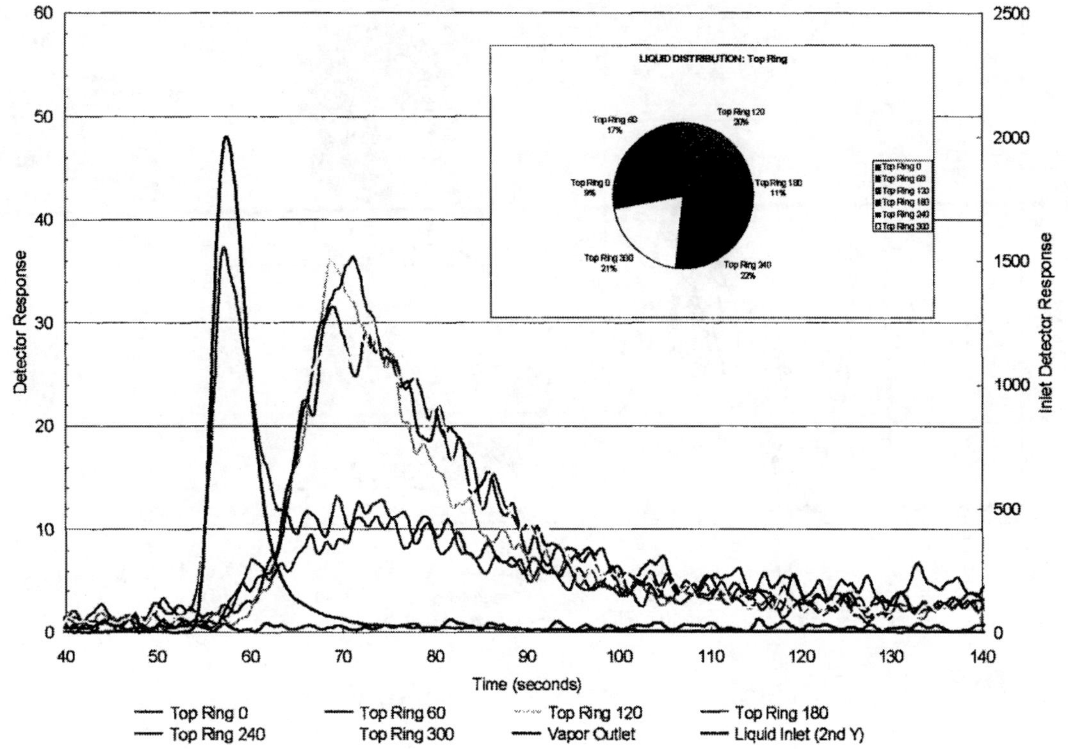

The analysis of the liquid distribution in the bottom ring of detectors showed preferential flow to the 120 and 240 degree quadrant with 22.1% and 20.3% of the flow, respectively. The 0, 60, 180, and 300 degree quadrants showed 13.5%, 15.2%, 14.1%, and 14.7% respectively. See Figure 8 below for graphical representation of the response curves.

HOME TOWN REFINERY
Distribution Study on the Main Column

Figure 8

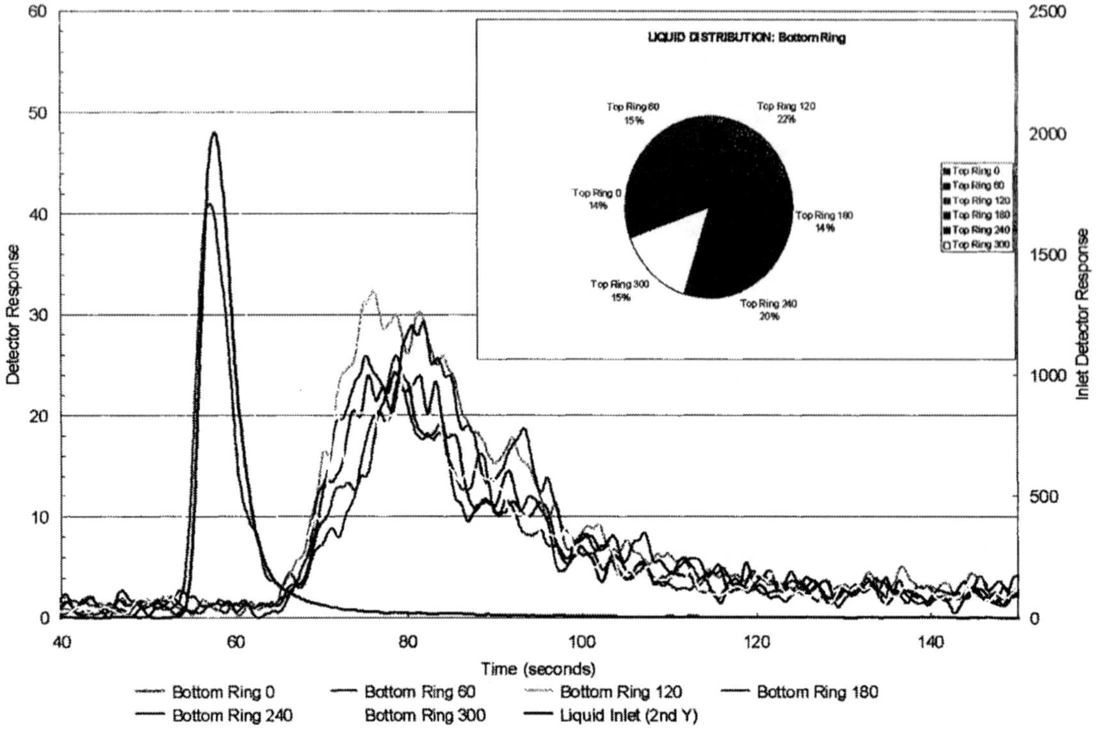

Conclusion

Theoretically, the vapor and liquid should behave the same in all quadrants, meaning the pulse profiles and timing should be identical for all. The data collected showed some preferential vapor flow to the 0 and 300 degree quadrants within the tower. The vapor rise velocities through the pack bed clearly indicate the areas of greater vapor flow. The 0 and 300 degree quadrants velocities were well above 4ft/sec while the opposing quadrants of 120 and 180 were less that 3ft/sec. The liquid distribution shows preferential flow to the opposing quadrants of 120 and 240 degrees. The 240 degree quadrant is interesting in that the percentage distribution is above expected level in both the liquid and vapor phases of the testing This is most likely caused by the change in vapor and liquid velocities in this quadrant that is symptomatic of significant back mixing due to maldistribution.

The information gained for this study aided in the evaluation of the vapor distributor and the trough type liquid distributor used in the tower. As expected, large amounts of catalyst were deposited in the trough distributor from startup causing the liquid distribution flow imbalance. The flow imbalance also caused coke buildup in the low flow areas further compounding the problem over time.

Gamma Ray Transmission Limitations in Packed Bed Towers

The application of gamma ray transmission technology is very effective in many situations. In the case of trayed towers a number of operational problems can be detected. However the technology must be used with caution in the case of packed bed systems. Usually in these vessels a gamma ray grid scan is performed using four equal length chords. The chords are selected to run through the bed such that two intersect the other two at an angle of 90 degrees. This arrangement allows the measurement of the presence and location of the distributor system, liquid height within the distributor and hence determines if the hardware is working correctly, is overflowing or is angled causing maldistribution of the liquids. In addition this strategy is normally able to detect position of the bed and significant blockages within as the density difference is large between vapor and solids. However when the actual problem is one of liquid or gas maldistribution the limitations of the technology must be understood by all involved in order that measurements obtained are interpreted in an appropriate manner and additional technologies are used to further explore possible sources of the problem.

In order to demonstrate the limitations of a grid scan when looking for liquid and gas maldistribution assume a standard packed bed tower with an average density of 15 lbs/cft and a wetted ratio of 10%. If the tower is running with good distribution the expected total wetted density will be 16.5 lbs/cft. In the case of a 50% liquid maldistribution of liquid within the packed bed the change in effective density will either be 17.25 lbs/cft or 15.75 lbs/cft. If it is assumed that the bed is 100 inches in diameter, then the change in density required to be detected will be 0.75 lbs/cft over 100 inches, or a radiation reduction of 1.15. If the bed was running normally the radiation signal would be influenced by 100 inches of density 16.5 lbs/cft. This results in a radiation reduction of 1.26 times. The increased reduction due to the maldistribution is calculated to be an additional 5% of the total reduction at normal conditions. Therefore, in summary, a 50% liquid maldistribution within a 100 inch diameter tower with a packing density of 15 lbs/cft and a wetted ratio of 1:10 will reduce the radiation signal at the detector by 5%.

It is theoretically possible to measure a 5% variation in radiation counts at a detector as long as all other measurement variables remain constant. However, looking at the practicalities of scanning a tower, a number of none ideal conditions occur. These conditions can affect path length between the source and detector due to either tower physical features such as strengthening rings or insulation supports or environmental influences such as wind. These can be corrected to a certain extent by guide ropes. However even with guide ropes small differences in path length will occur during data gathering.

To demonstrate the effect of path length, a 5% reduction in radiation intensity will be generated by a change in path length between the source and detector of 2.5 inches across a 100 inch tower. That is, if the source and detector each move by more than 1.25 inches the response at the detector will be the same as a 50% liquid maldistribution event occurring within the tower. In a typical tower scanning situation such as a 100 feet vessel, it does not take significant source or detector movement to generate a 1.25 inch change in both the source and detector. Figure 9 below illustrates how a small change in distance may mask actual distribution trends.

Figure 9

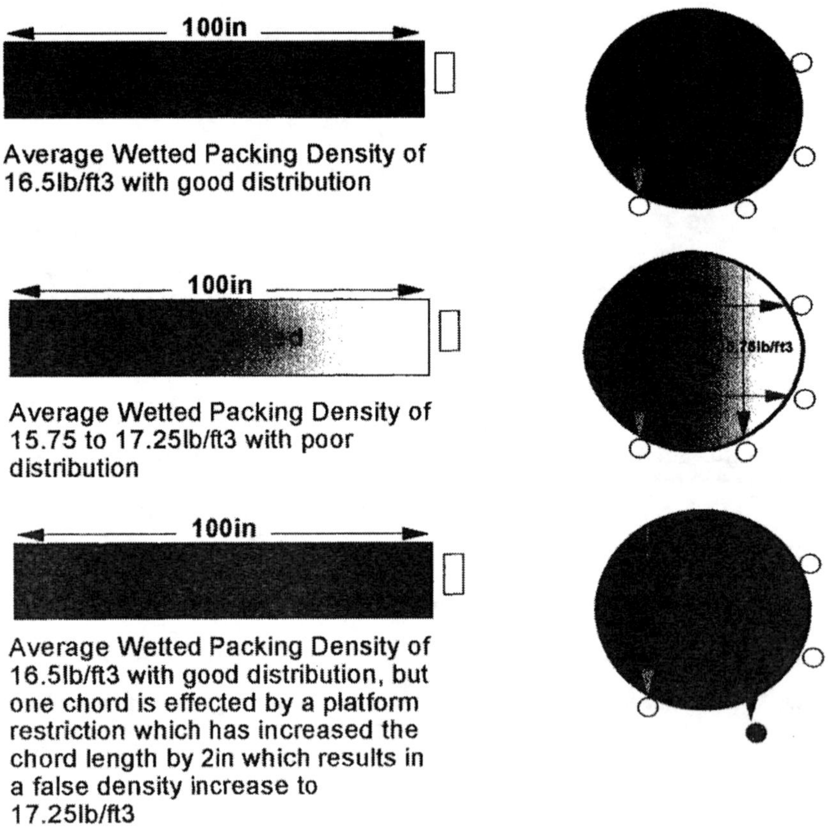

Average Wetted Packing Density of 16.5lb/ft3 with good distribution

Average Wetted Packing Density of 15.75 to 17.25lb/ft3 with poor distribution

Average Wetted Packing Density of 16.5lb/ft3 with good distribution, but one chord is effected by a platform restriction which has increased the chord length by 2in which results in a false density increase to 17.25lb/ft3

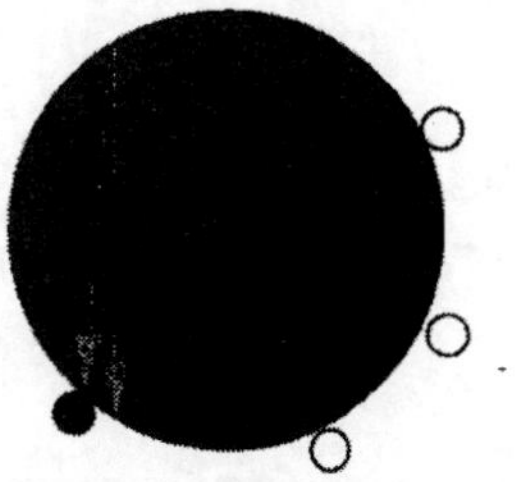

Average Wetted Packing Density of 16.5lb/ft3 with good distributio, but one chord has been affected by wind and has moved 1.5 inches closer to the source causing the density to decrease to 15.7lb/ft3

It must also be stressed that path length is only one factor affecting the accuracy of data collection. One other significant parameter is the horizontal movement of source and detector. Clearly when scanning through a chord of a cylindrical vessel any movement horizontally outward will increase the effective wall thickness. A 5% decrease in radiation intensity that is the same response as would be registered by a 50% maldistribution of liquid with the bed, can be achieved by a wall thickness change of only 2/100 th's of an inch on each wall. In many thick walled vessels a 2/100th change can occur with minimal horizontal movement.

One other point should be stressed when evaluating packed bed densities, is the variation typically observed in radiation counting. Two practical considerations have to be addressed. The first consideration is the potential statistical variations in radiation counting. This number is typically twice the square root of the count rate. This variation can best be illustrated by figure 10. Figure 10 shows the statistical variation in radiation counting varies the density calculation by as much as +/- .8lb/ft3 in the dry packing from the above 100 inch diameter example and +/- .9lb/ft3 in the wetted packing. The graph clearly shows an overlapping area that could confuse the analysis of the data.

POTENTIAL STATISTICAL VARIENCES IN RADATION COUNTING

Figure 10

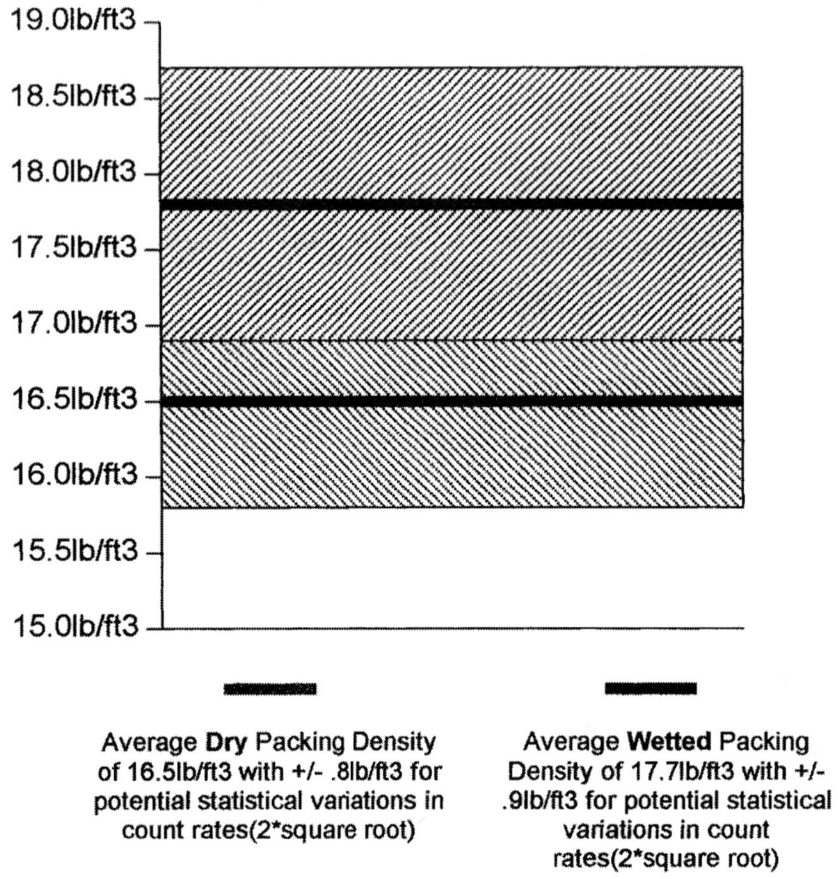

The second practical consideration in the radiation source sizing needed to get good statistical data. As the radiation intensity increases through the tower and packed bed, the better the statistical averages become. Example: at the lowest amount of radiation detectable above natural background radiation or 0.03 mR/hr, the statistical variation in terms of density would be +/- 2.2lb/ft3. As the radiation intensity increases the variations becomes less evident, but as the diameter of the tower becomes larger then radiation source size needed becomes an issue or in cases of large diameter towers in excess of 18ft almost impractical. The table below illustrates the tower diameter versus source size needed to obtain good statistical data.

Tower	Activity Needed(mCi)
48"	1
96"	27
144"	290
192"	2808
240"	16260

It can therefore be concluded that a 50% maldistribution of liquid within a packed bed is simply not detectable with any confidence using gamma ray transmission technology. In operational terms this level of maldistribution would have a profound effect on tower efficiency.

Therefore when applying radioisotope technologies to a packed bed tower it is recommended that the following strategy is used:-

1. Gamma Ray Transmission Scan Using One Chord of a Grid Scan Set. This will allow very obvious problems to be detected with minimal costs such as a missing distributor, missing or partially missing packed bed or flooded tower.
2. Gamma ray Transmission Gamma Grid Scan to assess the presence of distributor problems or localized significant blockages within the packed bed.
3. Injection of Compatible liquid and gaseous radiotracers into the process with rings of detectors located at critical positions around the bed to detect the presence and extent of liquid or gas maldistribution.

A case study showing the effectiveness of this strategy is given below:-

Necessity for Radioisotope Technology Integration to Fully Understand Packed Bed Operations

A major LNG supplier based in the Americas was experiencing excessive carryover of liquid from a packed bed tower. When the tower was operated at 100% design rate, liquid carryover of between 0.1 and 1.5 percent of the total liquid traffic was observed. When testing at 75% of the full rates, the carryover was reduced in proportion to volumetric flow. The customer requested a suitable radioisotope study to be designed and implemented to diagnose the source of the carry over problem.

The initial approach was to perform gamma ray transmission scan on the system to ensure that the demister pad, distributor and packed bed were located in the correct positions and functioning correctly. The scan was also to be used to determine any gross solids build up, liquid hold up or gross liquid maldistribution within the bed.

Gamma Ray Transmission Scan Results

In general, the vessel internals appeared mechanically intact with no evidence of flooding. No evidence of damage to the vane-pack demister was identified. The upper (liquid) distributor tray was in place and appeared to be holding approximately 6 to 8 inches of liquid. The bottom of the packed bed was measured at the design elevation indicating no bed slumping. The top of the packed bed was located approximately 12 to 14 inches below it's design elevation, suggesting either settling, loss of packing, or that the bed was never originally filled to it's design volume.

A comparison of the wetted packing density through various scan chords showed slightly decreased liquid holdup in the West or Northwest section of the top one third of the bed. Bed densities, i.e. liquid holdup, appeared uniform in the bottom two thirds of the bed. The lower (vapor) distributor was in place at the design elevation.

The four chord scans at the lower process rates are shown below in figure 11. This is shown as a relative density scale versus elevation

Figure 11

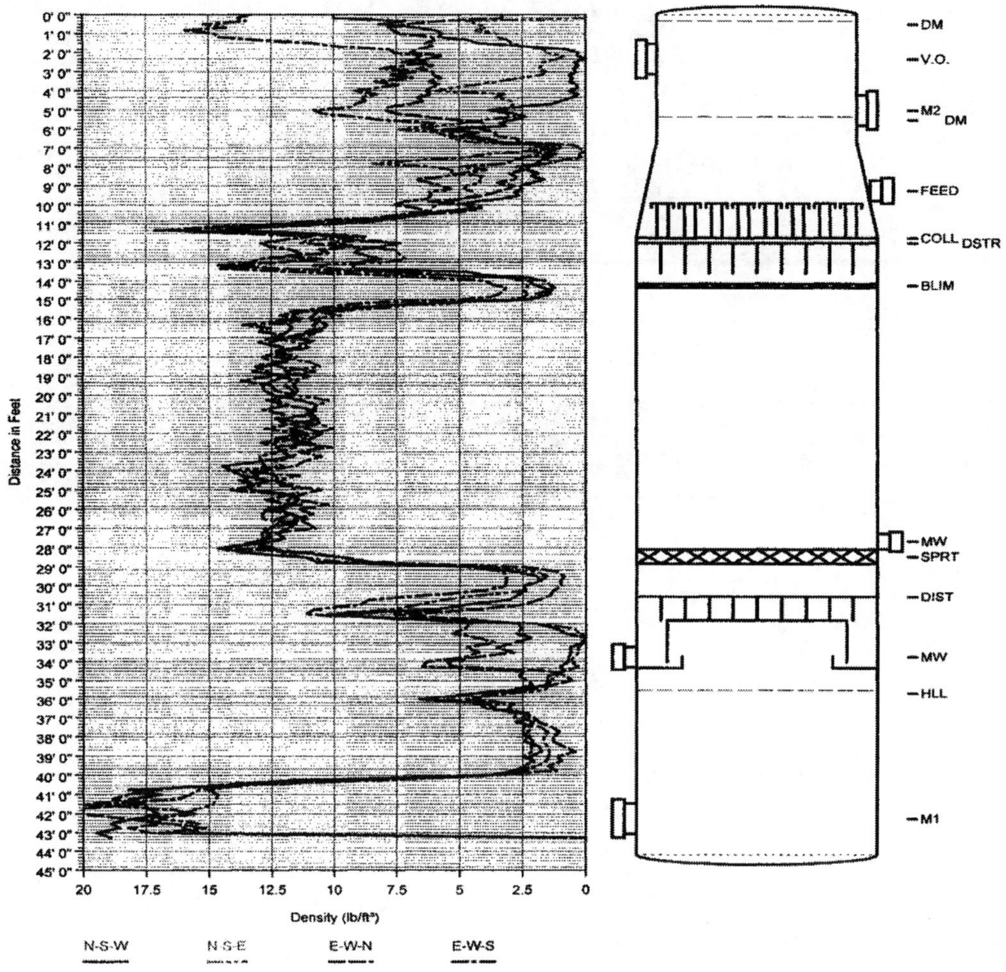

Radiotracer Distribution Studies

As no obvious problem was determined through use of the gamma ray transmission technology it was decided to use radiotracer technology to determine distribution of the liquid component in the feed and the vapor component in the stripping gas in order to identify flow distribution and residence time of the two components through the vessel.

Detectors were placed at vertical intervals down the vessel, around the bed (three rings of four detectors), at the vapor exit point and around the liquid and vapor distributors. In total 25 detectors were deployed. Elevations of the detectors are shown schematically in Figure 12.

Figure 12

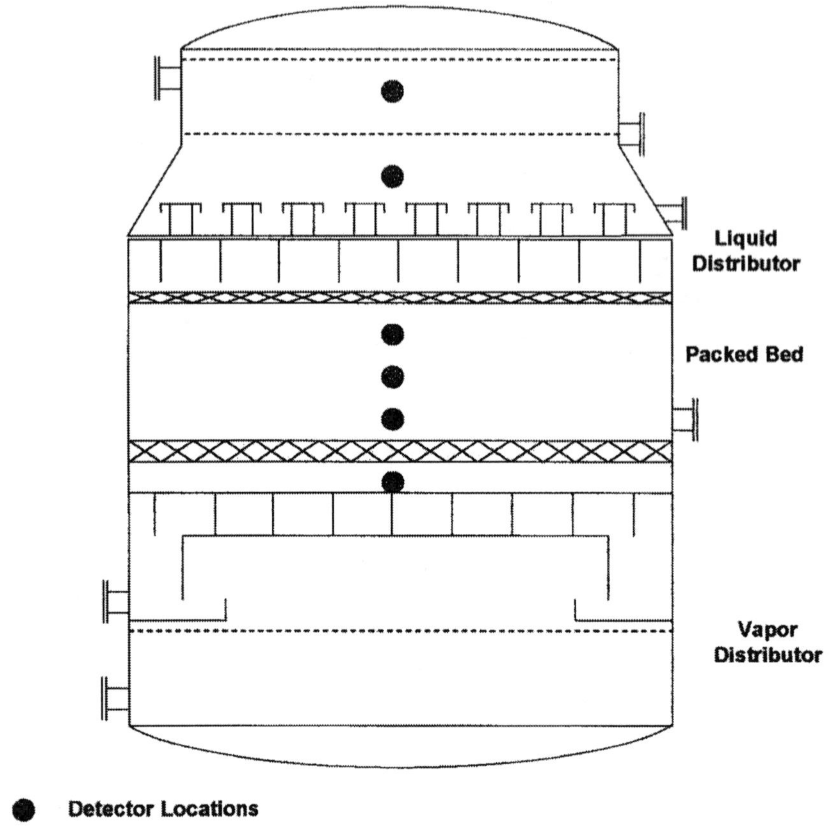

Injection of compatible liquid tracer was carried out at five different operating conditions. The feed inlet deflector plate appeared to split the liquid fairly evenly to both sides of the vessel. The liquid distributor appeared to uniformly distribute the liquid to the bed as in all cases the radial flow distribution was uniform. The observed liquid distribution in the four quadrants both at the distributor and within the bed was approximately 25% for each quadrant (+/- 1%) indicating no liquid maldistribution.

At the vapor distributor elevation some liquid flow imbalance was measured with more liquid (up to 30% in total) flowing down the Northwest side of the tower. Lowering the feed rates and repeating the test improved this to approximately 27%.

However the main observation from the liquid tracer test was the variation in liquid velocity down the bed that was detected at all rates. The velocity down the South was much faster than down the North quadrant ranging from 0.29 ft/sec to 0.63 ft/sec.

.

Injection of vapor tracer was carried out at three process conditions into the stripping gas rather than the feed gas.

In all cases a severe vapor bias was observed within the bed. Although the stripping gas appeared to be uniformly distributed by the vapor distributor, all of the vapor traffic appeared to travel up the Southwest, Northwest, and Northeast sides of the vessel. Very little vapor traffic traveled up the Southeast side. When the vapor reached the top of the bed, it was determined that 35% of the total traffic was diverted down through the Southeast quadrant of the bed. Upon exiting below the bed it followed the original flow path of the stripping gas and the cycle repeated with diminishing amounts of recycle gas with each recycle pass. This resulted in a vapor mean residence time within the vessel in excess of 3 minutes.

Figure 13 shows the response from detectors at the bottom of the bed with flow only up three sides of the tower. The small response from the Southeast detector was due primarily to tracer presence flowing within the other quadrants of the tower.

Figure 13

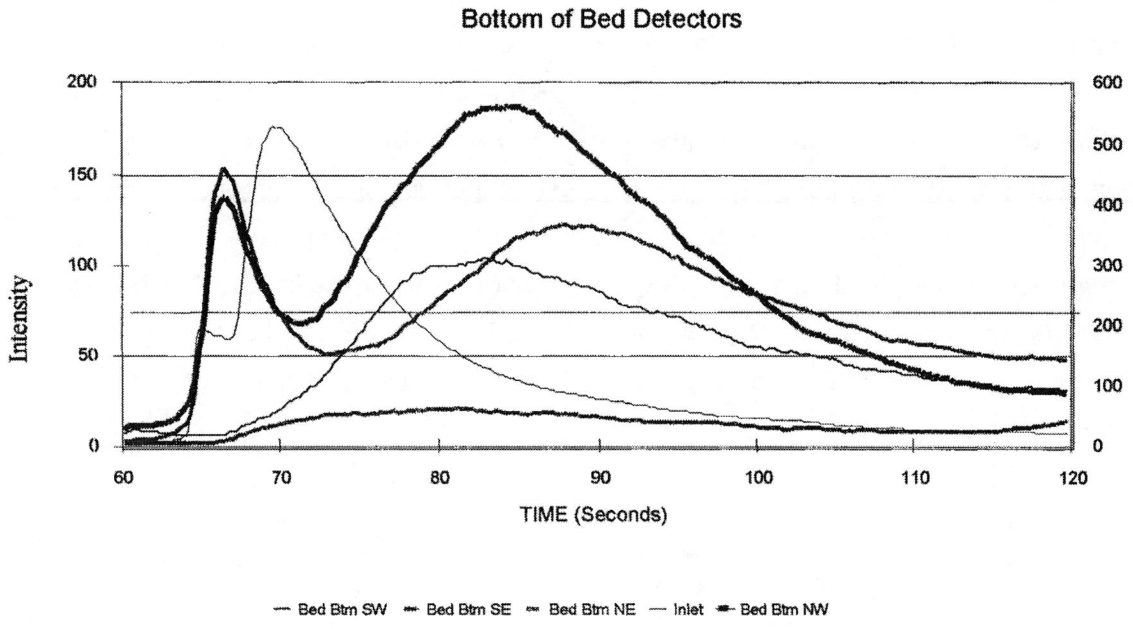

Figure 14

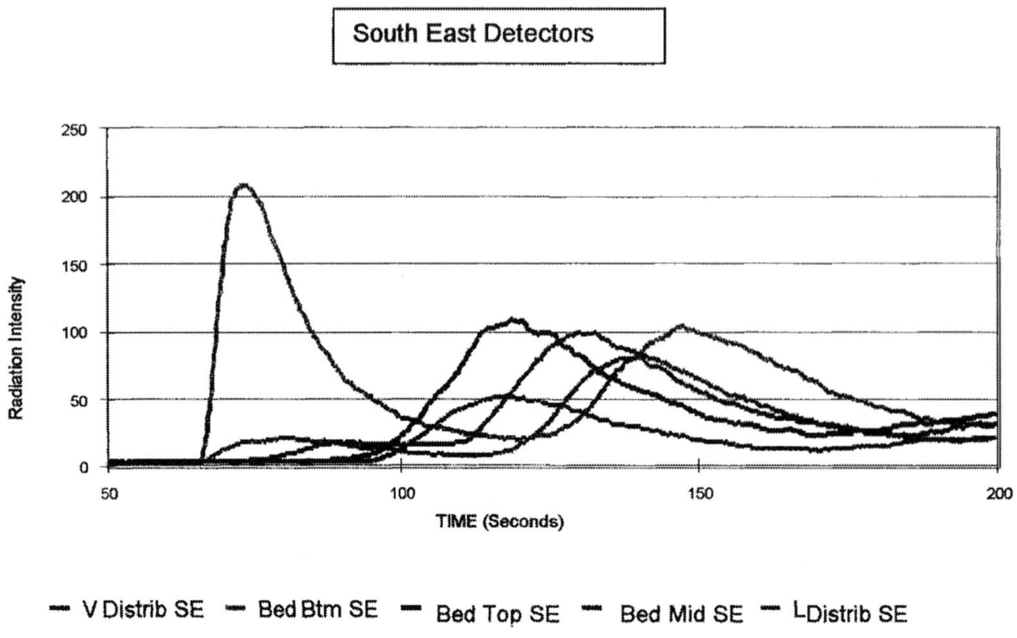

Figure 14 above shows the Southeast quadrant of the tower. The detector monitoring the vapor distributor showed the first response, (as part of the initial distribution), then the next detector response came from the detector monitoring the top of the bed. The detectors then responded downward in sequence: top of the bed, middle of the bed, bottom of the bed, and then again at the vapor distributor

The final figure 15 below shows the response from detectors located in the Northwest quadrant of the tower. For comparison, the following figure shows the responses of all detectors located on the Northwest quadrant of the vessel. Note that the detector response sequence indicates a flow up the vessel, responding in the sequence: vapor distributor, bottom of bed, middle of bed, and top of bed. To support the recycle measurement in the Southeast quadrant, the "up-flow" responses in the Northwest quadrant occur at an earlier timing than the "down-flow" responses observed on the Southeast side of the vessel. The secondary responses from the recycled gas can the also be observed.

Figure 15

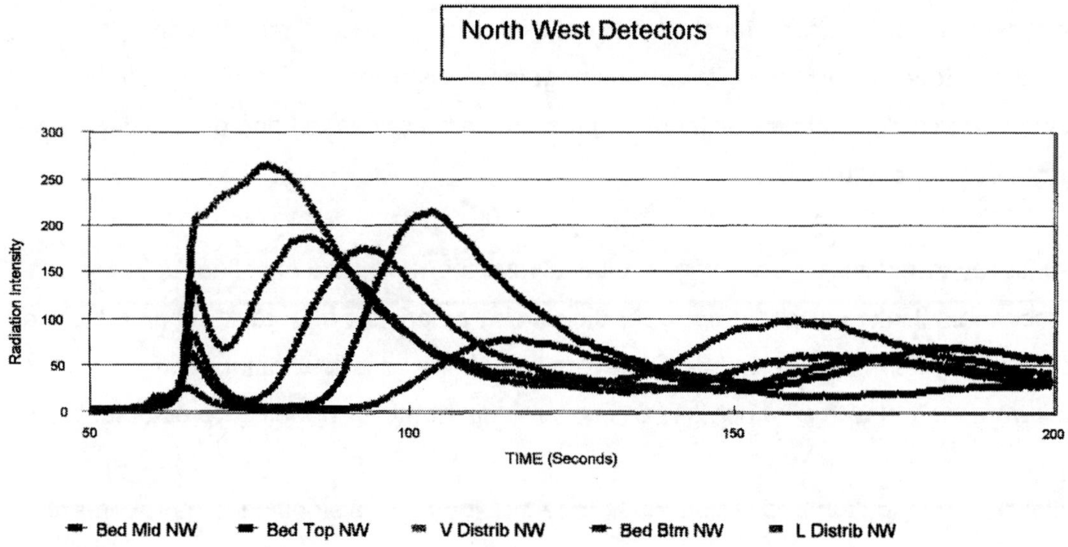

Summary

No mechanical or flow distribution problems were observed using a gamma ray transmission quadrant scan technique. However a small liquid and significant vapor maldistribution and recycling flow problem was measured at a variety of conditions through use of a liquid and vapor radiotracer study.

The gas recycle within the system is a relatively uncommon occurrence and the tests that were performed did not identify the cause of the recycle. The onsite engineering consulting team's speculation as to the cause of the recycle included an ice accumulation in the bed. However, if significant ice had accumulated in the bed, its presence should have been identified using the gamma ray transmission scanning technology.

Given that flow rate of the vapor component of the feed was roughly an order of magnitude greater than the stripping gas flow rate, it appears possible that the stripping gas recycle issues could be the result of a feed gas mal-distribution problem. At the time of publishing this paper no tests have been carried out to measure distribution of the feed gas within the system.

Conclusion

It is therefore concluded that radioisotope technologies can be very useful tools in diagnosing operating characteristics of process vessels including both trayed and packed bed towers. However, limitations of gamma ray transmission must be fully understood in order that flow distribution problems are not missed and discounted as the source of tower malfunction.

Gamma ray transmission scanning is relatively inexpensive, and excellent at verifying the mechanical integrity and gross operational characteristics of a packed bed tower and distribution system. However, it faces severe limitations as a flow distribution measurement tool with as much as 50% flow maldistribution not being detectable.

Radiotracer distribution and residence time technology offers significant improvements in flow distribution measurement sensitivity, provides a measurement of traffic residence time, and allows distribution measurement of both liquid and vapor traffic to be determined.

In all packed bed diagnostic studies careful planning must be undertaken that uses gamma ray transmission as well as neutron backscatter to detect gross mechanical problems within the tower. Radiotracer use must be addressed as the second stage of the study if no obvious problems are determined using the scanning techniques. This approach ensures that all possible fluid flow distribution issues are covered as part of the tower performance investigation.

ADVANCED HYDRAULIC AND MASS TRANSFER MODELS FOR DISTILLATION COLUMN OPTIMIZATION AND DESIGN

R. Bruce Eldridge, Christian Green
The Separations Research Program
The University of Texas

Valmor de Almeida
Oak Ridge National Laboratory

Introduction

The U.S. petroleum refining and chemical processing industries consume over 12 quadrillion BTUs of energy each year. Distillation is a low thermal efficiency unit operation (about 6% for easy separations) that currently accounts for 40% of the processing energy consumed in refining and continuous chemical processes.[6] In spite of the high energy required for distillation, it is often chosen over other separations processes because of the relatively low initial capital investment, flexibility, and ability to yield high purity products. Currently, every barrel of crude oil is subjected to an initial separation by distillation, and nearly every chemical process requires distillation for product recovery or purification.[19] Because of high initial capital costs, alternative technologies are not likely to be implemented until existing distillation columns are replaced. By contrast, substantial savings in energy consumption can be achieved by relatively small process efficiency improvements in distillation column operation and column internal design. This research project develops an advanced hydraulic and mass transfer model and validates it at commercial scale. The model can then be used to improve column performance and thus reduce the large energy consumption in distillation operations in the chemical and petroleum refining industries.

Lack of understanding of the mass transfer and flow dynamics occurring between phases inside distillation systems is the primary barrier to improving their energy and process efficiencies. Current hydraulic and mass transfer design models[9,10,13,16] are based on limited empirical measurements and visual observation. These models incorporate variables derived from overall performance of the entire column (composition profiles, exit stream concentrations, pressure drop, gross liquid holdup, temperature profiles, etc.). Predictive models based upon local conditions inside the column are needed to set operating conditions of high performance distillation columns and design column internals that produce optimum performance.

High performance packings that have low liquid holdup, low pressure drop, and high flow capacity have been developed over the last twenty years for use in distillation columns. They yield lower energy requirements than more traditional packings for a given amount of mass transfer between the liquid and gas streams. Despite the improvements in packed tower design, the development of packing elements remains an empirical art. Scientific based design is expected to be the next development stage that brings distillation to a higher level of performance.

Project Overview

The project involves the development of a CFD-based model for the hydraulics of structured packing. X-ray tomography and fundamental experiments are being conducted to assist in the development and validation of the CFD model. Results to date have been very promising.

CFD Model Development

Computational results (Figure 1) have been compared with data obtained from a dual plate experimental system (Figure 2). The experimental apparatus permits high-fidelity representation of its geometry. Results of average linear pressure drop from the comparative simulation agree very well (within 5%) with the corresponding experimental data. The gas flow factor range in the simulation spans realistic limits as required by the industrial participants in the project (0 to 2.5 ft/s $(\text{lb/ft}^3)^{0.5}$). The calculations employ an 8 CPU cluster and entail the solution of the equations of motion for a Newtonian fluid (Navier-Stokes equations) via GAMBIT/FLUENT software.

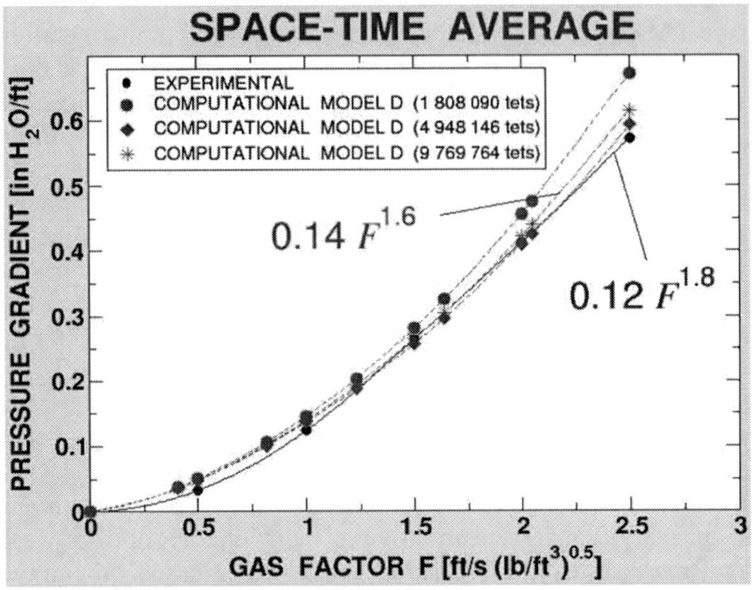

Figure 1 – CFD results versus experimental dry pressure drop

Figure 2 – Dual plate apparatus

In extending the model development to realistic packing elements, an automated and systematic interface was built and named GraSPI (Graphical Structured Packing Interface – Figure 3). The interface is but one deliverable of the project that allows industrial users to quickly assemble realistic packing elements and perform a single-phase flow simulation. The interface was built as a plug-in for the GAMBIT/FLUENT software (Fluent Inc.). Fluent Inc. is a participant in the project and GraSPI could turn into a commercial product. The current version of GraSPI has a built-in library of commercial packing elements, and automatic problem setup for users unfamiliar with computational fluid mechanics. The interface creates the packing structure, defines the mathematical domain, and outputs scripts to drive the underlying solution software with minimum intervention from the user. As a bonus, a blue print of the geometry may be delivered for fabrication purposes.

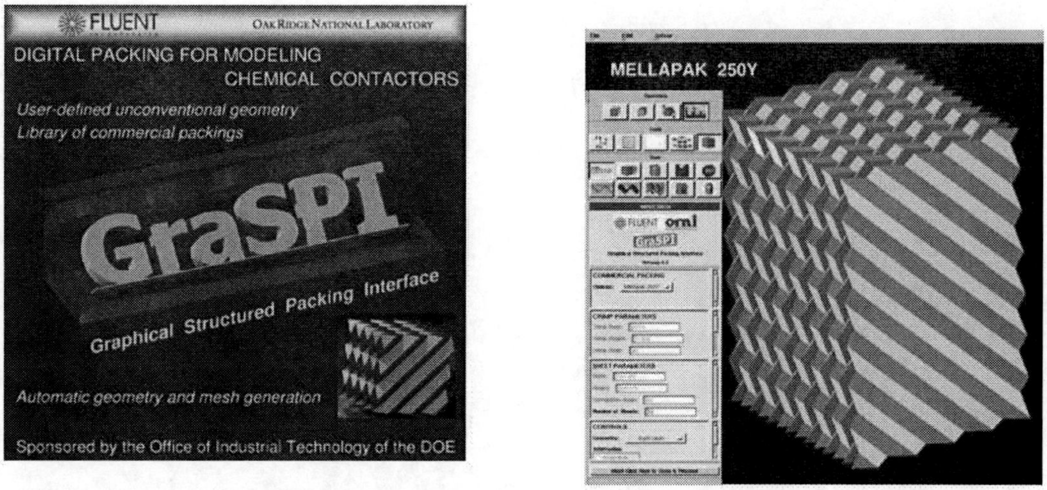

Figure 3 – GraSPI user interface

GraSPI has been tested and used for analysis of typical commercial packings used in distillation columns. Recently a joint effort with the participants in the project (Dow Chemical, Fluent Inc, Sulzer Chemtech, and SRP UT-Austin) led to preliminary results from simulations using realistic packing geometry. Three packings (Figure 4) were used, namely, Mellapak 205Y, Montz B1-250Y, and Montz B1-250X, for the calculation of linear pressure drop and associated flow quantities, such as flow trajectories, kinetic energy, and velocity magnitude. Packing blocks with 10 and 14 sheets were used with simulations performed on an 8 CPU cluster.

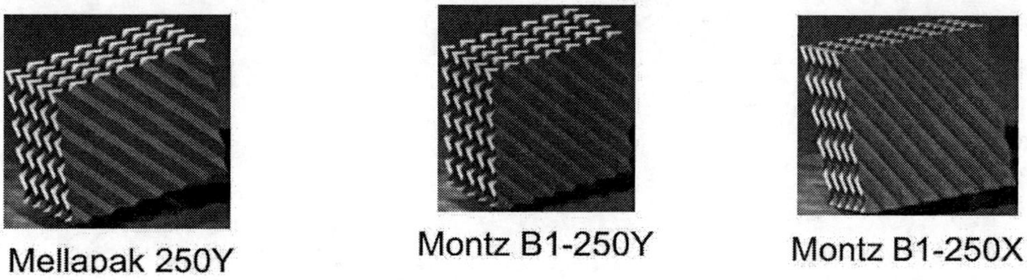

Figure 4 – Packings used for implementation of GraSPI

One aspect of the simulation was to assess the hardware investment industries need to make in order to take advantage of current computational fluid dynamic codes. Our conclusion so far is that clusters equipped with at least 10 CPU's are necessary for resolution of flow scales in one single packing element (the above packing elements are 7.5-in tall). The simulations have been suggesting interesting insights into the flow field not accessible to experimental methods at this point. For instance, paths of individual fluid particles can be followed and local pressure drop evaluated as they zig-zag between the corrugated sheets (Figure 5). It was found that local pressure drop could be as high as 3 times the average between the ends of the packing element. More importantly, comparison with SRP experimental data for the Montz B1-250Y in a packed column proved very encouraging (Figure 6). .

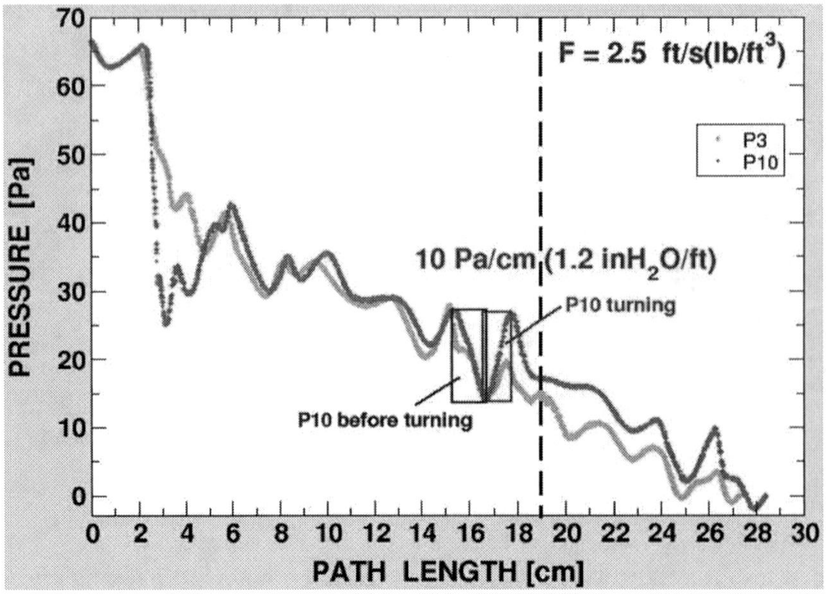

Figure 5 – Localized pressure profile

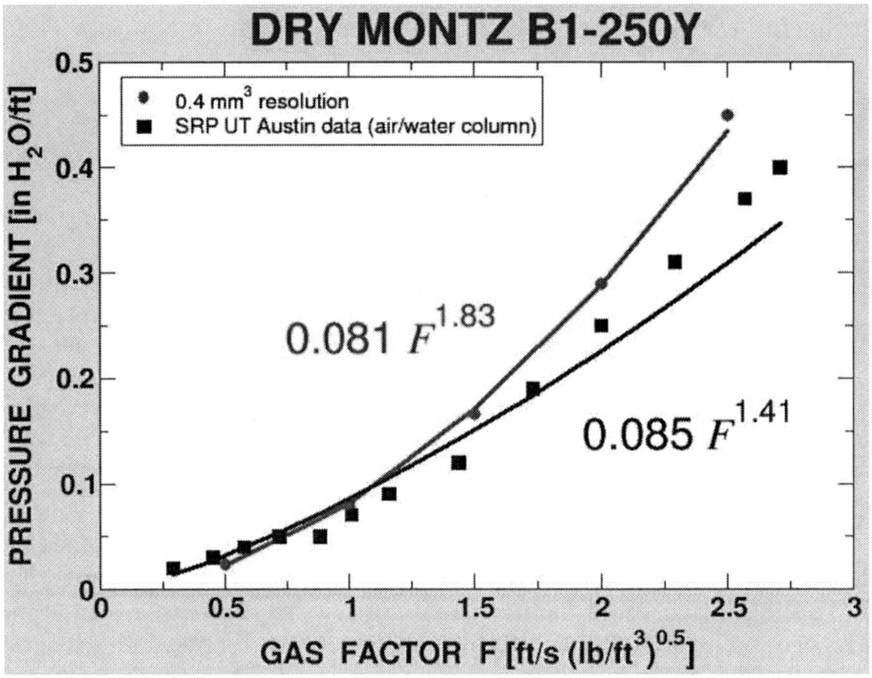

Figure 6 – CFD predicted dry pressure drop versus SRP column results

Future work

GraSPI is currently being enhanced to include perforations on the corrugated sheets, and to generate cylindrical packing elements suitable for stacking and filling of cylindrical vessels such as distillation columns (Figure 7). In addition, the spacing between the packing elements and the column wall is under implementation. Gas-liquid flow simulations are also in development. The first test case is gas-liquid counterflow through a pair of corrugated plates of the experimental setup described in the previous section; experimental data has been collected for validation. Other cases with packing elements will follow.

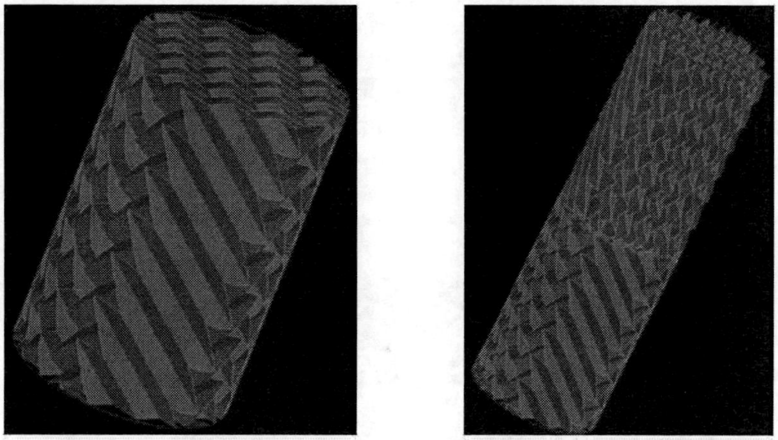

Figure 7 – GraSPI representation of cylindrical packing elements

With results achieved so far there is a possibility of extending GraSPI to the larger scope of modeling flow through a pilot distillation column of the size of the air/water column at SRP-UT Austin. That column is equipped with 16 packing elements of 7.5-in height each and 16-in diameter, resulting in elements with 34 corrugated sheets (Figure 8).

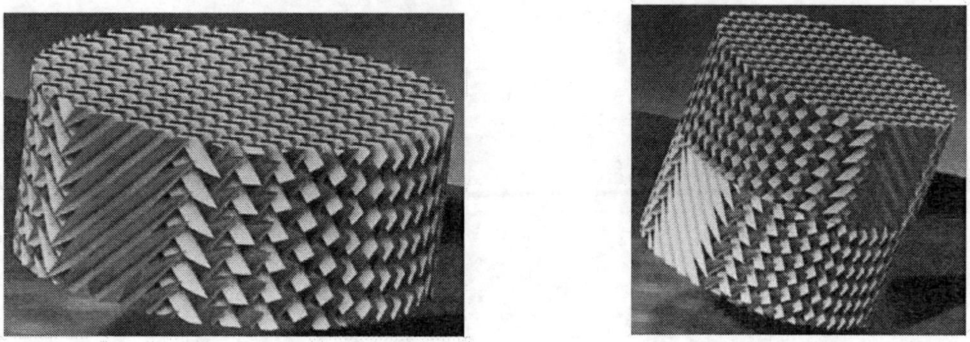

Figure 8 – GraSPI representation of SRP pilot plant packing elements

Simulations of this magnitude will involve 100 to 1000 CPU's and would represent a major step toward first-principles modeling of flow through distillation towers. The significance of this development is that US chemical industries would have access to a modeling tool with an unprecedented level of detail. This tool could be used in a variety of ways. Notably, the tool could be used for designing improved packings which could then be shipped to the SRP for testing under design conditions. Alternatively, industrial users of the tool could have the SRP test packing elements used to calibrate the tool for a proprietary scale-up design. In any scenario, the tool would satisfy the purpose of reducing energy consumption in the distillation industry through state-of-the-art design. Furthermore the tool could become a design standard in the distillation allied industry, and also be used for educating the future generation of engineers.

X-ray Imaging

The X-ray CT System used in the project is shown in Figure 9. The X-ray source generates a fan beam of X-rays which pass through the column and impact the detector array. The intensity of the X-rays incident on the detector is a direct measurement of the density along the X-ray path. A reconstruction algorithm is applied to results obtained from a 360 degree rotation of the column. The result is a slice of the column cross-section (Figure 10)

Figure 9 – GE X-ray CT system with column in place on turntable

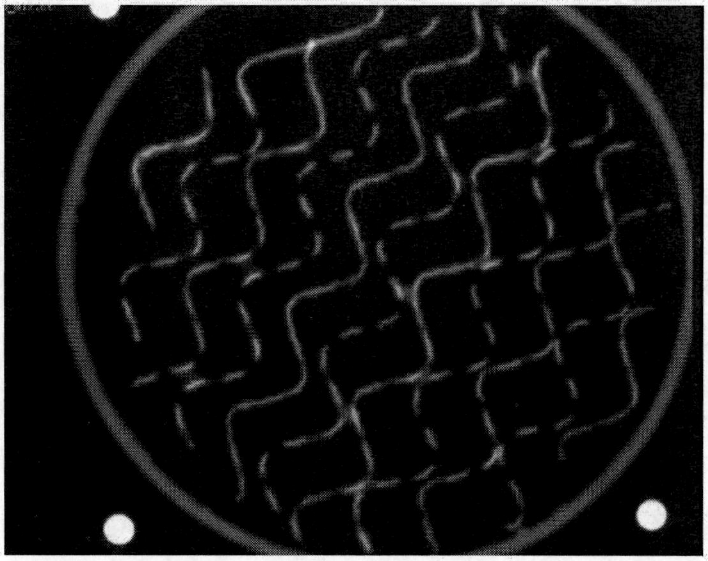

Figure 10 - Reconstructed slice of column

A close-up of the surface of a packing element of Mellapak 250 Y is show in Figure 11. The liquid films are clearly visible. A 3D scan composed of sequential slices through a 3 inch high section of the packing element is show in Figure 12. Areas of dry packing are shown in the upper right corner of the image. These results indicate that a very high percentage of the packing surface is covered by liquid despite the relatively high surface tension of water.

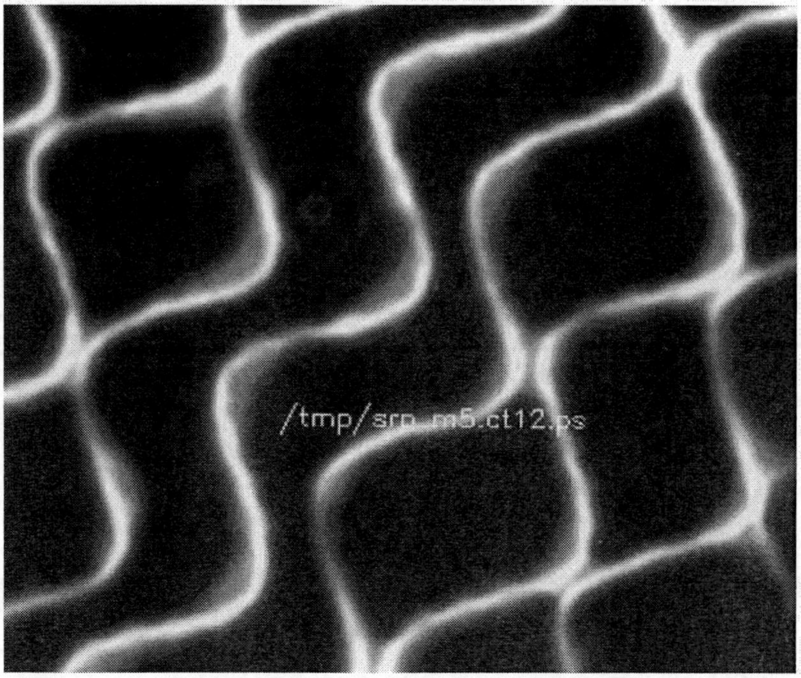

Figure 11 – X-ray scan of Mellapak 250 Y

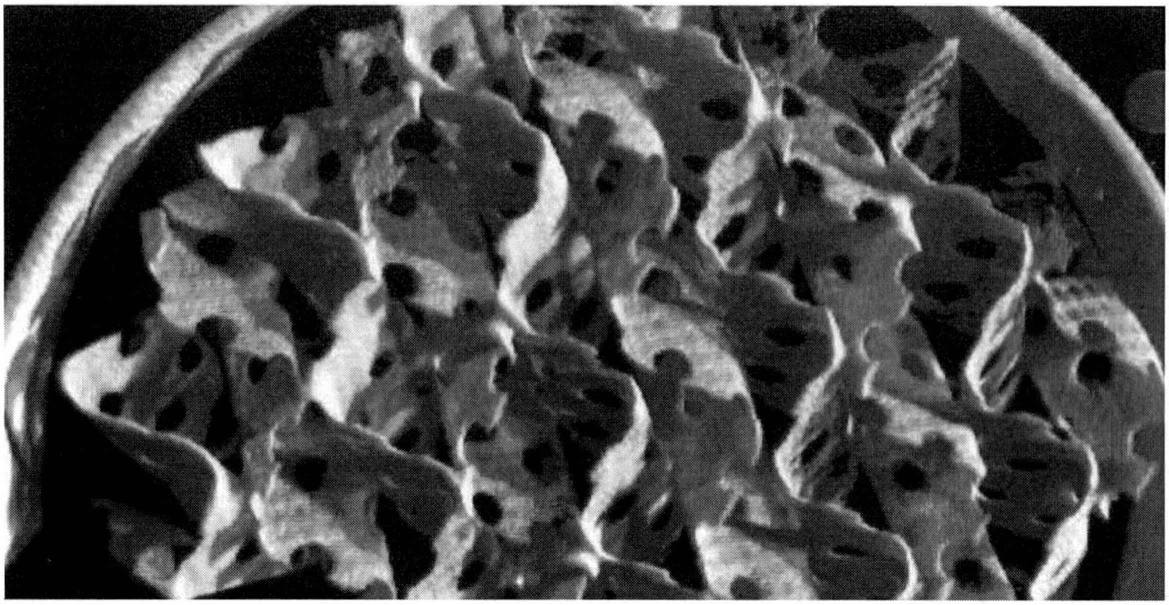

Figure 12 – 3D Scan of 3 inch section of Mellapak 250 Y

Future Work

The X-ray CT scans will be analyzed and the liquid hold-up and wetted surface area will be quantified. Additional scan analysis techniques will be used to extract additional data from the single slice and 3D images.

References

1. Cartmel, DB. MSE Thesis, University of Texas at Austin, 1999.
2. Roadmap for Computational Chemistry, Council for Chemical Research, 2000.
3. Computational Fluid Dynamics Technology Roadmap, 1998.
4. *Emissions of Greenhouse Gases in the United States 1996.* DOE/EIA-0573(96), Energy Information Administration, Office of Integrated Analysis and Forecasting, US DOE, Washington, DC, October 1997.
5. *Energy and Environmental Profile of the U.S. Petroleum Refining Industry.* U.S. DOE/Office of Industrial Technologies (1998).
6. Humphrey, JL and GE Keller. *Separation Process Technology.* McGraw-Hill, New York (1997).
7. Maier, RS. *Phys. Fluids* 10 (1), 60-74, January 1998.
8. Fair, JR. Personal communication. July 1999.
9. Rocha, JA, JL Bravo, and JR Fair. *Industrial & Engineering Chemistry Research* 32 (4), 641-651 (1993).
10. Rocha, JA, JL Bravo, and JR Fair. *Industrial & Engineering Chemistry Research* 35 (5), 1660-1667 (1996).

11. Schmitz, D, N Reinecke, G Petritsch, and D Mewes. *Conference Proceedings: Frontiers in Industrial Process Tomography II*, 103-108 (1997).
12. *Separation Technologies for the Industries for the Future,* National Research Council, National Academy Press, Washington, DC (1998).
13. Stichlmair, J, JL Bravo, and JR Fair. *Gas Separation & Purification* 3, 19-28 (1989).
14. Toye, D, P Marchot, M Crine, and G L'Homme. *Chemical Engineering Science* 49 (24B), 5271-5280 (1994).
15. *Vision 2020: 1998 Separations Roadmap.* Center for Waste Reduction Technologies (AICHE), New York (1998).
16. Wagner, W, J Stichlmair, and JR Fair. *Industrial & Engineering Chemistry Research*, 36, 227-237 (1997).
17. Zarabi, T and A Kantzas. *Canadian Journal of Chemical Engineering* 76, 853-865 (1998).
18. Darton, R.C., *I. Chem. E. Symp. Ser.*, 128, A385-90, 1992.
19. *Technology Roadmap for the Petroleum Industry*, draft report, February 2000.
20. *Office of Industrial Technologies: Enhancing the Competitiveness, Efficiency, and Environmental Quality of American Industry Through Technology Partnerships*, U.S. Department of Energy, 1997.
21. Porter, K. E. *Trans. IChemE.*, 73, 357-62, 1995.
22. Humphrey, J. L. Separation Technologies: Marketing Factors. Final Report, DOE/ID/12920-2, NTIS DE92014524, 1991.
23. de Almeida, V. F. and Riahi-Nezhad, C. K. ORNL Report ORNL/TM-2000-101, in press in *Sep. Sci. Technology 2000.*

Superstructure formulation for batch distillation synthesis

Santosh Jain, and Robin Smith
Department of Process Integration,
UMIST, PO Box 88, Manchester, M60 1QD, UK

Abstract

A synthesis framework is presented for screening of different configurations and sequences for batch distillation processes. A superstructure is generated using a generic unit, which consists of segments of trays with top and bottom vessels attached to it. Three different types of connections; feed connections, sequencing connections and configuration connections, generate the superstructure using the generic unit. The superstructure is formulated as a mixed integer nonlinear programming (MINLP) problem. A hybrid optimisation method using simulated annealing and sequential quadratic programming (SA-SQP) is used to solve the superstructure. A case study is presented to show the application of the approach.

Introduction

Batch distillation is a unit operation frequently used in fine and speciality chemicals. The transient nature of batch distillation allows it to operate and be configured in different ways. The batch rectifier is the most common configuration. Novel configurations are also explored due to increased interest in other batch distillation arrangements. The inverted column (batch stripper), middle vessel column, multi-vessel column are considered novel configurations. Along with the different configurations and sequences, batch distillation units can be operated in different manners. General operating policies are, constant reflux-reboil policy in which product composition varies throughout the operation, and variable reflux-reboil policy which keeps product composition constant. However, the most profitable operation is to operate the column at optimum reflux-reboil policy. In the multi-vessel and the middle vessel columns, distribution of the feed provides another degree of freedom for the optimal operation.

To synthesise a batch distillation the flow-sheet decisions are made for the selection of best configuration, or to arrange the best sequence. Due to the development of novel configurations, and different operating policies, the batch distillation synthesis is not a trivial task. General practice in screening of configurations is based on heuristics. Kim and Diwekar (2000) presented some heuristics rules for the selection of the best configuration. Heuristics were developed with the results of many simulation runs for three component mixtures of different compositions. Heuristic rules usually contradict, and are applicable for extreme cases not for intermediate cases. Moreover, conditions of the column change, so the heuristic which are applicable at the start of the operation may not be applicable further in the operation.

Development of the novel configurations allows more options for selecting the best configuration. Also, a batch distillation flow-sheet may include more than one configuration. A systematic way is needed to screen different configurations, and different sequencing combinations. Previously, Sundaram and Evans (1993) solved the batch distillation synthesis problem considering only the batch rectifier. A network of batch rectifiers was proposed allowing the sloppy separation to provide maximum throughput. Oldenburg *et al.* (2001) considered batch distillation sequencing problem for a batch rectifier-stripper combination. The framework developed by them considers rectifier-stripper combinations. They used a distillation unit consisting of top and bottom vessels, so that a column can be operated either as a rectifier or as a stripper. In this work a similar distillation unit is used, called generic unit.

Previous work for batch distillation synthesis is limited to certain configurations. In this work a more general superstructure is proposed for the batch distillation synthesis. The superstructure is formulated as an MINLP problem, which is solved using a SA-SQP algorithm, and a set of batch distillation networks is proposed.

Synthesis approach

The proposed system representation accounts for different configurations, sequences, and operating policies. A generic superstructure is proposed. The generic unit of the superstructure is a tray segment with top and bottom vessels attached to it. The superstructure includes different configurations, and different sequencing of them. Figure 1 shows a superstructure proposed for three products. The superstructure consists of three types of connections: (i) feed connection, which distributes the feed amongst the different vessels, (ii) configuration connection, which combines different segments to form a new configuration, and (iii) sequencing connection, which connects different segments amongst each other, and to the products.

A middle vessel column and a multi-vessel column can be generated by combining different generic unit. While combining two units, at the junction point bottom vessel of top unit is merged with the top vessel of bottom unit. The superstructure also allows options for vapour to pass or to bypass intermediate vessels. Such configurations are generated using configuration connections. If the units are arranged in a sequence, the superstructure allows the opportunity to pass the intermediate feed from one column to other columns in different ways. Intermediate feed can be charged to a single vessel of the unit, or can be distributed among the different vessels of the same unit. Sequencing connections allow generation of such different sequences. Sequencing connections also allow the separation task to be chosen amongst the tasks available to the unit.

Model of batch distillation unit

The dynamics of batch distillation has been analysed by rigorous stage-to-stage models to shortcut models. Diwekar (1995) provides details of different models available for batch distillation units.

In this work a semi-rigorous model is used to analyse the dynamics of the column. The compartment analysis, based on the work by Benallou (1986) is used to develop a semi-rigorous model. Figure 2 shows a compartment of a distillation section. The compartment analysis considers cumulative holdup of each tray at one sensitive tray. A differential mass balance equation is set-up across the sensitive tray, and an algebraic mass balance equation is set up across other trays. Liquid holdups of the trays are assumed to be constant. Vapour holdup is neglected. Condensers and reboilers are considered separate compartments. The energy balance is not considered in the model. Ideal stages are assumed, so that the vapour and liquid leaving a stage are in equilibrium. Pressure drop is neglected in the column.

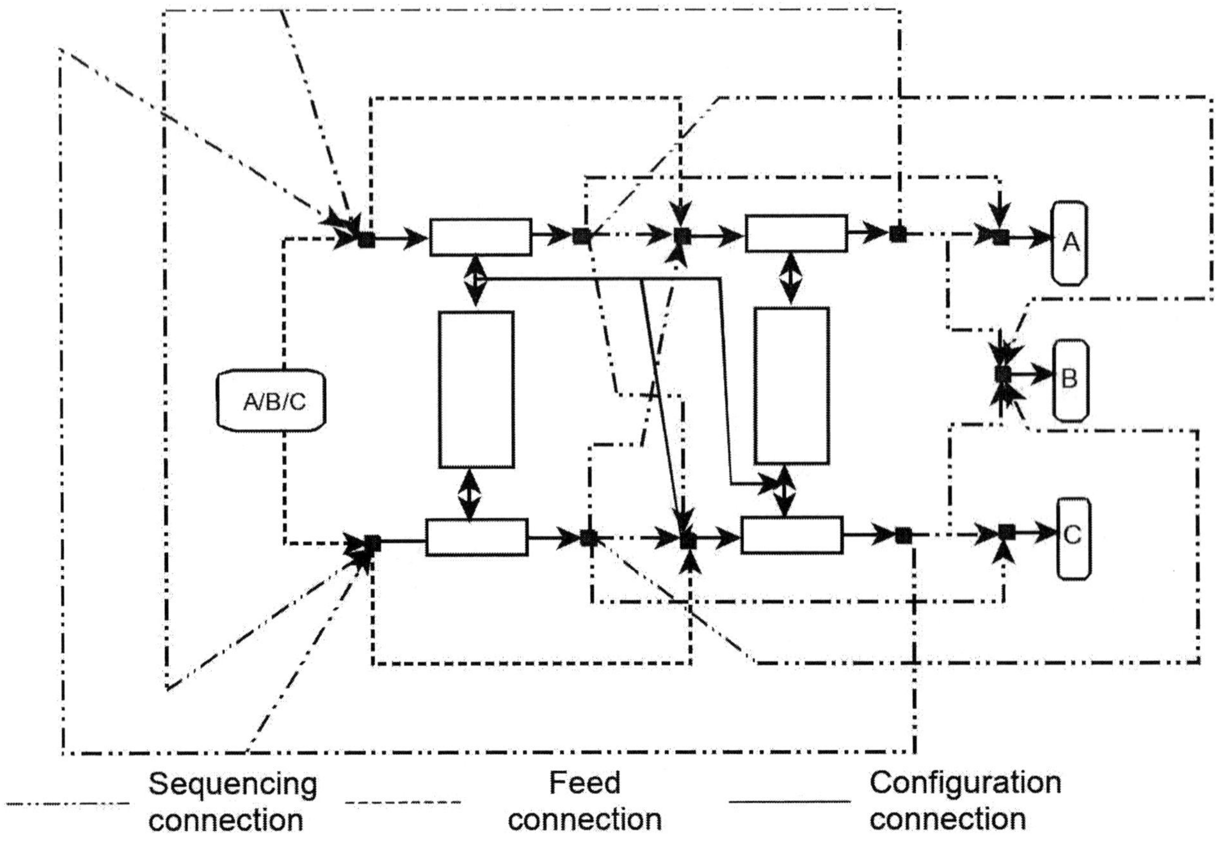

Figure 1 Superstructure representation for batch distillation systems

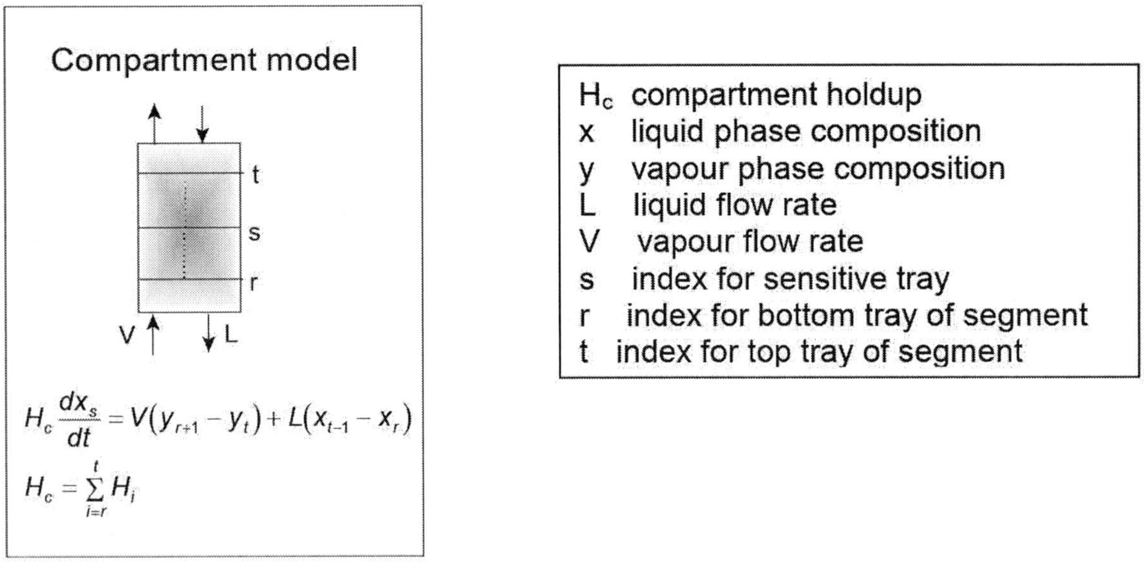

Figure 2 Compartmental model for batch distillation column

Optimal operation of batch distillation column

Control variables in the batch distillation column are the reflux ratio, the vapour flow rate, product flow rate. For the optimal operation of a column the control variables should follow the optimal profiles. The optimal operation of a batch distillation has been studied by various researchers. Mujtaba and Macchietto (1996), Logsdon and Biegler (1993), Li *et al.* (1998) used rigorous model and control vector parameterization for optimal operation of batch rectification. Furlonge *et al.* (1999) developed a rigorous model for the multi-vessel column and studied the optimal operation at different modes. Farhat *et al.* (1990) converted a dynamic optimisation problem into a nonlinear programming problem using a linear and an exponential reflux policy for batch rectification.

In this work, the control variable follows a profile. Shape, size and value of profile are considered the optimisation variables. Applicability of the proposed approach depends on the versatility of the profile generation algorithm. Choong and Smith (2004a) developed a novel profile generation tool, which can generate a wide range of practically possible profiles. Six parameters are required to target a profile. The proposed algorithm to generate profiles contains two basic kinds of profiles, exponential and asymptotic profiles. With the combination of these two profiles, a wide range of profiles can be generated. They used the profile generation tool for the optimisation of batch crystallisation processes. In this work the same profile generation tool is used for the control variable parameterization.

Feasibility of batch distillation column

Success of the synthesis approach depends upon the procedure for extraction of the feasible flow-sheets and its operation. For the batch distillation column, feasibility is characterized by the desired purity of the products. Selection of the operating policy and the design to provide a feasible design and operation is not a trivial task in batch distillation process. Previous studies in batch distillation have shown that the synthesis, and optimization of batch distillation always faces this problem. Also, during the optimization, the most of the effort is wasted to determine the feasible operation and design. Moreover, due to the complex nature of the model (rigorous model, and semi-rigorous model), it becomes time consuming.

Diwekar (1992) used a shortcut model for optimization of batch rectifiers, and proposed feasibility criteria for the batch distillation. For a path constrained variable, it is easy to meet the constraints in comparison to an end point constrained variable. This is because the value of the end point constrained variable is only available at the end of the period. Choong and Smith (2004 a, b) used profile generation tool to meet the path constraints. They used the approach, to avoid the dry-up problem in the batch crystallisation unit by selecting the holdup of the solvent as the control variable. They applied a profile generation algorithm for the control variable, and imposed bounds on the profile parameters, to meet the path constraints. For endpoint constraints it is not possible. Using the profile generation tool, bounds can be set at the endpoint but the feasible bounds are not available for intermediate values.

In this work we propose a *limiting gradient* approach. Figure 3 shows how the approach works. If at any point T_o the value of the endpoint constrained variable is X_o, and at endpoint T_f the value is X_f, then limiting gradient is calculated according to the Figure 3. So, at any point t the limiting gradient is calculated, and it is multiplied by a factor greater than 1 (safety factor) to obtain the new gradient at time t. The multiplication factor is taken as the optimisation variable (controlled variable). With the calculated gradient, operating variables are determined using corresponding equations. The concept behind the approach is to achieve the target value before the endpoint. Also the assumption is made that a potential available at point T to meet the target is greater than the next point. So extra work can be done at the beginning rather than later. If it is done later a penalty will have to be paid.

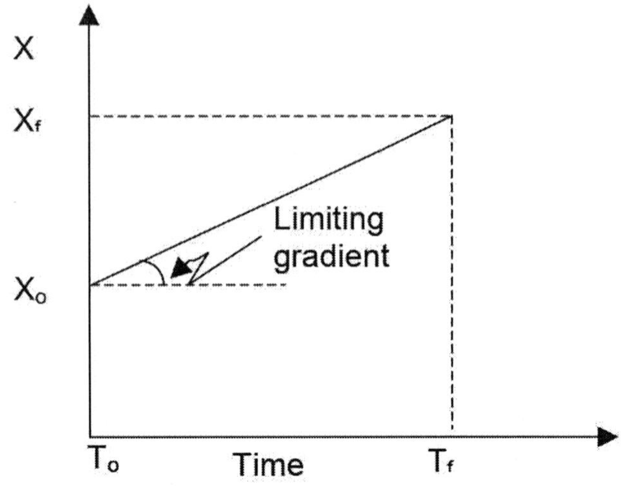

Figure 3 Limiting gradient for endpoint constrained variable

For the batch distillation endpoint constrained variables are chosen as product purity and recovery. Using the above approach operating variables (liquid and vapour flow rates) are calculated.

Solution method

The superstructure discussed so far is formulated as an MINLP problem. The problem is solved using a SA-SQP hybrid approach. Figure 4 shows how the approach works. Success of simulated annealing depends upon the cooling schedule and Markov chain length at any temperature level. In the approach used here, simulated annealing provides the structural changes and initial feasible operation for the structure. With the structure and operation obtained from SA, SQP performs NLP and provides local optimal solution. The SQP also provides the long Markov chain length for SA. Termination criterion is checked by SA, which is either the final temperature or the maximum number of perturbations made by SA. It is possible to run SQP multiple times for the same structure for the same or different initial operating conditions. It eventually provides multiple initial points for SQP, and helps to determine the global optima.

Case study

A case study is performed to demonstrate the applicability of the approach. The case study is taken from Noda *et al.* (2001). The objective function is performance index (P.I.) as defined below,

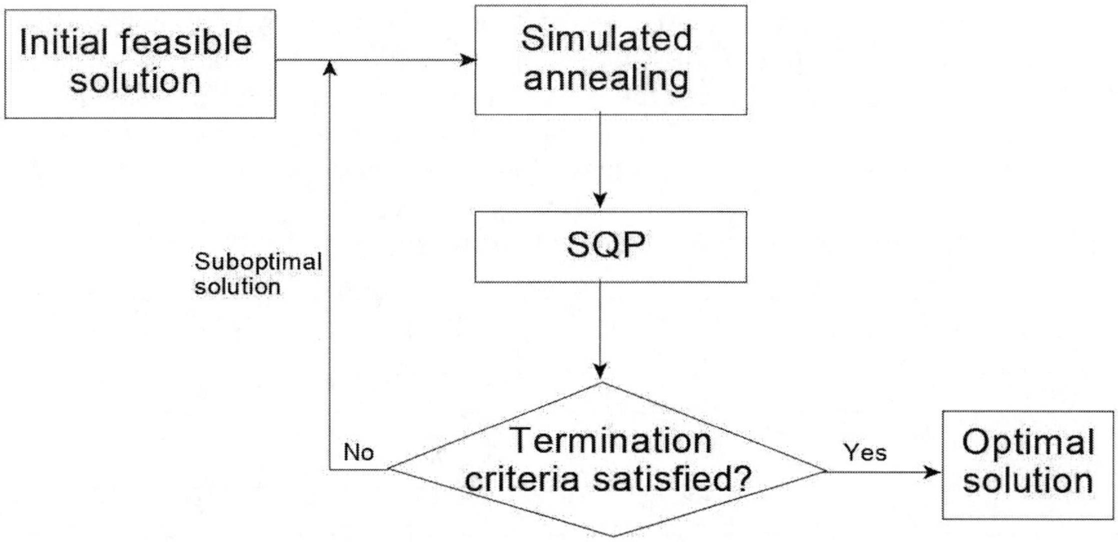

Figure 4 Optimisation strategy for the superstructure solution

$$P.I. = \frac{total\ products\ (mol)}{batch\ time\ (h)}$$

Table 1 Column data

Total number of stages	10
Feed	103 kmol
Vapour flow rate	50.0 kmol/h
Liquid flow rate	20 – 50 kmol/h (rectifier) 50- 80 kmol/h (stripper)
Number of components	3 (A/B/C)
Products	3
By products	2

Table 2 Feed and product data

	A	B	C
Feed	0.3	0.4	0.3
Product 1	0.95		
Product 2		0.95	
Product 3			0.95
Relative volatility	9.0	3.0	1.0

Problem data are given in Table 1 and Table 2. The initial structure provided to the optimization is sequence of four batch rectifier, using a single unit. Table 3 provides the set of solutions obtained from the optimization. A sequence of a rectifier followed by a stripper is obtained with the significant improvement from the conventional rectifier sequence.

A middle vessel performs the best amongst all the configurations and sequences for the given problem. Next in the order is rectifier-stripper with feed distribution. In this sequence feed is initially distributed in the rectifier, *i.e.* part of the feed (29%) is given at the top of the column, and column is operated at total reflux. Product A is taken from the top of the rectifier and product B and C obtained from the stripper. The same column is operated in reverse after the rectifier operation; the charge from the bottom of the rectifier is now fed to the top of the column, in this way column operates as a stripper. The same rectifier-stripper combinations, without feed distribution in rectifier also provide near equal performance with the previous case. For the only rectifier and only stripper sequence byproducts are allowed to be taken. The conclusion can be made that if the single column with one segment has the flexibility to operate it either as a rectifier or as a stripper, it can perform near to the performance of middle vessel. Note should be taken here, that time for the charge from one vessel to other vessel is not considered, which might affect the ranking of the solutions obtained.

Table 3 Different structures obtained from the optimization

	Middle vessel	R-S (feed distributed)	R-S	Rectifier (R)	Stripper (S)
Time (h)	2.43	2.45	2.5	3.4	6.2
By-products (kmol)	0.0	0.0	0.0	8.5	21.8
P.I.	42.4	41.5	51	27.8	13.0

Conclusion

A novel approach is proposed for the synthesis of batch distillation processes. A superstructure is generated using a generic unit. With the combination of three different types of connections (feed, configuration, and sequencing) different flow-sheets for batch distillation can be obtained. A batch distillation unit is modelled using the semi-rigorous model based on the compartmental approach. A new feasibility criterion is proposed, for the feasible operation of the column. The superstructure is solved using hybrid optimization (SA-SQP) technique, and a case study demonstrates the applicability of the approach.

References

Benallou, A., D. E. Seborg, and D. A. Mellichamp, "Dynamic compartmental models for separation processes," *AIChE J.,* **32**, 1067 (1986).

Choong, K. L., and R. Smith, "Optimization of batch cooling crystallization," *Chem. Eng. Sci.,* **59**, 313 (2004 a).

Choong, K. L., and R. Smith, "Novel strategies for optimization of batch, semi-batch and heating/cooling evaporative crystallization," *Chem. Eng. Sci.,* **59**, 329 (2004 b).

Diwekar, U.M., "Unified approach to solving optimal design-control problems in batch distillation," *AIChE J.,* **38**, 1551 (1992).

Diwekar, U.M., *Batch Distillation: Simulation, Optimal Design and Control,* Taylor and Francis, Washington, (1995).

Farhat, S., M. Czernicki, L. Pibouleau, and S. Domenech, "Optimisation of Multiple-Fraction Batch Distillation by Nonlinear Programming," *AIChE J.*, **36**, 1349 (1990).

Furlonge, H. I., C. C. Pantelides, and E. Sorensen, "Optimal operation of multi-vessel batch distillation columns," *AIChE J.*, **45**, 781 (1999).

Kim, K. J., and U. M. Diwekar, "Comparing batch column configurations: Parametric study involving multiple objectives," *AIChE J.*, **46**, 2475 (2000).

Li, P., H. A. Garcia, G. Wozny, and E. Reuter, "Optimization of a semibatch distillation process with model validation on the industrial site," *Ind. Eng. Chem. Res.*, **37**, 1341 (1998).

Logsdon, J. S., and L. T. Biegler, "Accurate Determination of Optimal Reflux Policies for the Maximum Distillate Problem in Batch Distillation," *Ind. Eng. Chem. Res.*, **32**, 692 (1993).

Mujtaba, I. M., and S.Macchietto, "Simultaneous Optimisation of Design and Operation of Multi-Component Batch Distillation Column- Single and Multiple Separation Duties," *J. Process Control,* **6**, 27 (1996).

Noda, M., A. Kato, T. Chida, S. Hasebe, and I. Hashimoto, "Optimal Structure and On-Line Optimal Operation of Batch Distillation Column," *Comput. Chem. Eng.*, **25**, 109 (2001).

Oldenburg, J., W. Marquardt, D. Heinz, and D. Leineweber, "Mixed logic dynamic optimization applied to configuration and sequencing of batch distillation processes," *ESCAPE 12*, Elsevier, Amsterdam, 283 (2002).

Sundaram, S., and L. B. Evans, "Synthesis of Separations by Batch Distillation," *Ind. Eng. Chem. Res.*, **32**, 500 (1993).

Generalized SVRC-QSPR Model for Prediction of Saturated Vapor Pressures and Phase Densities

S. S. Godavarthy, D. Ravindranath, R. L. Robinson, Jr., and K. A. M. Gasem
School of Chemical Engineering, Oklahoma State University,
Stillwater, OK

Abstract

In response to the stringent economic and environmental standards, the chemical industry has been exploring new ways to develop compounds. These compounds should seamlessly combine cost considerations, environmental fate and process safety, in addition to desirable physical, chemical and thermodynamic attributes. Computer aided molecular design (CAMD) deals with development of such designer chemicals. Thousands of new molecules, most of whose properties are unknown, can be generated using genetic algorithms. In most cases, promising subsets of these molecules are synthesized and then tested for their properties. Clearly, this is an expensive and time-consuming approach. An alternative approach is to develop efficient and reliable property models based on the chemical structure of the molecules. In this work, we develop Quantitative-Structure-Property Relationships (QSPR) for predicting the parameters in our Scaled-Variable-Reduced-Coordinate (SVRC) framework. Specifically, we demonstrate the efficacy of the SVRC model in providing accurate saturation property predictions for a diverse molecular dataset when QSPR parameter generalizations are used. In general, the SVRC-QSPR model is capable of predicting the vapor pressure and phase densities for over 40 fluids with average errors of less than 1%.

Introduction

The chemical industry faces considerable economic and environmental challenges in the 21st century. Major forces for change in the industry include the demand for improved environmental performance, increased profitability, productivity, and changing work force requirements [1]. One of the important conclusions after years of research is that the growth and competitive advantage of the chemical industry would depend upon the rapid development of new methodologies to design novel chemicals. The classical approach of molecule design, involving synthesis, testing and validation is becoming obsolete due to the high costs associated with such a design scheme. Due to these problems, and rapid improvements in the computational power available to researchers, the industry has been working toward the development of a computer aided molecular design approach. One field that has seen wide spread use of "designer chemicals" is that of distillation. Distillation is by far the most widely used method in the chemical and petroleum industries, for production of high purity chemicals. Worldwide, about 95% of all chemical separations are done by distillation [2]. In the U.S. alone, some 40,000 distillation columns exist, accounting for a capital investment of about $8 billion. They require energy equivalent to about 15% of all U.S. industrial energy consumption. These statistics clearly illustrate that improvements in distillation technology can reap huge financial benefits [3].

Design and operation of processes using alternate molecules requires accurate knowledge of thermo-physical behavior of the chemicals. Experimental data on thermo-physical behavior is often limited or non-existent for new chemicals designed through genetic algorithms. The traditional approach in the industry for handling saturated phase properties has been to develop equations-of-state (EOS) and saturation property models. In previous work we developed the SVRC model [4] which was capable of predicting the saturation properties for over 40 molecules within 0.1% AAD with two regressed parameters per molecule. The generalization of the SVRC parameters provides accurate saturation property predictions when used to model normal fluids such as alkanes. However, when the same generalization approach was extended to highly polar fluids such as alcohols, the model predictions were not as accurate. An alternative to this approach has been the use of structure-property relationships to model properties based on chemical structure information. These QSPR models however do not incorporate any theoretical framework and have been limited to estimating properties at a single temperature and thereby were not typically used to correlate saturation properties over the full saturation range. In this work, we plan to integrate the SVRC and the QSPR models, where the SVRC model provides a theoretical framework for relating the effect of temperature on the property while the QSPR model provides accurate parameter generalizations based on chemical structure. Here, we attempt to model challenging molecules like refrigerants and aromatics. The results obtained indicate that this combined SVRC-QSPR model provides better prediction results than either the SVRC or the QSPR models used separately.

Previous Model Development

In our previous studies [4] the SVRC framework was used to correlate saturation properties of a wide variety of organic molecules over the entire saturation range. The SVRC model with two regressed parameters per molecule was able to represent vapor pressure of most molecules within 0.1%. Also the previous modeling efforts showed that the SVRC model has excellent generalized predictions of saturation properties with average errors of less than 1% based on available physical properties. The general SVRC formulation for correlating thermo-physical properties may be given as [5]:

$$P=\left\{P_c^{\alpha}-\left(P_c^{\alpha}-P_t^{\alpha}\right)\Theta\right\}^{1/\alpha} \tag{1}$$

Where $\Theta(\varepsilon)$ = correlating function, α = scaling exponent, T = temperature, P = saturation property at given T, and suffix c,t represent critical and triple points respectively. The values of $\Theta(\varepsilon)$ and α are dependent on the saturation property being calculated:

For vapor pressure

$$\Theta=\frac{1-A^{\varepsilon^{B}}}{1-A} \qquad \frac{\alpha_c-\alpha}{\alpha_c-\alpha_t}=\frac{\varepsilon+C\varepsilon^2}{1+C} \tag{2}$$

For liquid density

$$\Theta=\frac{1-A^{\varepsilon^{B}}}{1-A} \qquad \frac{\alpha_c-\alpha}{\alpha_c-\alpha_t}=\frac{1-A^{\varepsilon}}{1-A} \tag{3}$$

For vapor density

$$\Theta=\frac{2-A_1^{\varepsilon^{B_1}}-A_2^{\varepsilon^{B_2}}}{2-A_1-A_2} \qquad \frac{\alpha_c-\alpha}{\alpha_c-\alpha_t}=\frac{2-A_1^{\varepsilon^{C}}-A_2^{\varepsilon}}{2-A_1-A_2} \tag{4}$$

where A, B and C are correlation constants, α_c and α_t are the limiting values of α at the critical temperature and triple point temperatures, respectively. α_c and α_t have been found to be highly structure dependent [6,7]. In the present study, we generalize the SVRC model parameters using the QSPR approach to find the relationships between the intrinsic molecular structures and the SVRC parameters.

QSPR Model Development

Due to the computational intensity and the lack of a proper theoretical basis, calculation of thermo-physical properties from first principles is difficult to attain [8]. The QSPR approach has become the tool of choice for correlating thermo-physical properties of chemicals over the last decade. The structure of a chemical compound can be used to map it's chemical and physical properties. Researchers long have been concerned with decoding this molecular map in order to understand the relationship between the behavior of the chemical and the structure of a molecule. The QSPR approach often provides predictions for chemical and physical properties of as-yet-unmeasured or unknown compounds based on structure information. High quality predictions can be obtained using these descriptors since structure-property mapping is at an atomic level rather than at a functional group level. In the literature, all QSPR model development efforts usually begins with a homogeneous data set and naturally progress into a more diverse set after an initial model has been developed. In this work we generalize the SVRC parameters α_c and $\Delta\alpha$ seen in the above equations. The above equations reveal that knowledge of critical properties is essential for developing good saturation property correlations. QSPR based non-linear models have been previously developed and validated for prediction of critical properties [9].

Dataset: A database containing of over 40 pure fluids including alkanes, alcohols and refrigerants was used in this study. The data set used to test the ability of the model to predict saturation data contained over 5157 saturated vapor pressure data points, 5128 saturated liquid density and 5023 saturated vapor density data points. For vapor density only 25 compounds were considered due to limitations in experimental data. Also, most of the refrigerants studied here do not have data over the complete saturation range from the triple point to the critical point. In such cases, where the data did not extend to the triple point, the lowest available data point was used as the lower-temperature end point. Physical properties (specifically T_c, p_c, Z_c and ω) required to implement the model were obtained from DIPPR [10].

Structure Entry: The names of the individual molecules were stored in EXCEL. The two-dimensional structures were then generated using ChemDraw [11]. The structures were all exported into Chem 3D Ultra [11], which was then used to generate the 3D structures of the molecules. The structures were all optimized using the MOPAC [11] module available in Chem 3D Ultra. The pre-optimized structures were submitted to the AMPAC 6.0 [12] program for further geometry refinement and for calculation of molecular orbital parameters. The AM1 parameterizations were used to calculate the quantum-chemical molecular descriptors. The output files from AMPAC [12] serve as the input for commercial QSPR packages [13], which are used to generate the molecular descriptors.

Descriptor Generation: A commercial package [14, 15] was used for the calculation of a large selection of molecular descriptors including variety of constitutional, topological, geometric, and electrostatic descriptors. Following is a brief description of the various descriptor categories:

- Constitutional Descriptors: These simple descriptors reflect only the molecular composition of the compound without using the geometric or electronic structure of the molecule e.g., number of atoms, number of bonds, number of rings, and molecular weight.
- Topological Descriptors: These descriptors provide the atomic connectivity in the molecule. These include molecular connectivity indices, substructure counts, molecular weights, weighted paths, molecular distance edge descriptors, kappa indices, electro topological state indices, and many other graph invariants.
- Geometric Descriptors: These descriptors are calculated to encode the 3D aspects of the structures and include such descriptors as moments of inertia, solvent-accessible surface area, length-to-breadth ratios, shadow areas, and gravitational index.
- Electrostatic Descriptors: These descriptors are calculated to encode aspects of the structures that are related to the electrons. Examples of electronic descriptors include partial atomic charges, HOMO energies, LUMO energies, and dipole moment.

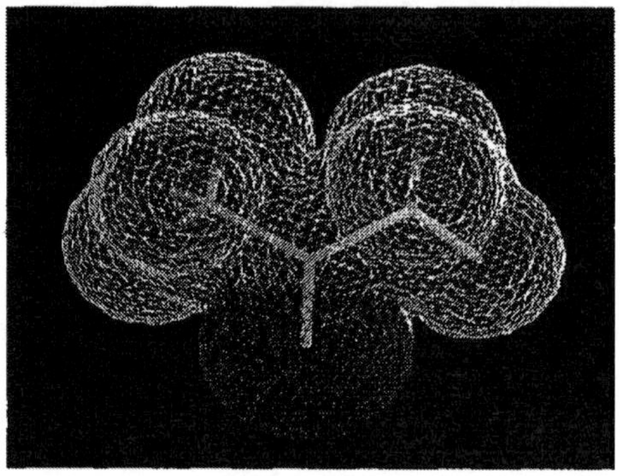

Fig. 1: 3-D Representation of Acetone

- Quantum Chemical Descriptors: These descriptors represent quantum-chemically calculated charge distribution in the molecules and, therefore, describe the polar interactions between molecules and their chemical reactivity. The descriptors also provide the value of the partial charge on the atoms in the molecule (e.g., dHmin represents the minimum partial charge on a hydrogen atom). Also, these descriptors relate to the strength of intra-molecular interactions and characterize the stability of the molecules, their conformational flexibility, and other valency-related properties, such as the maximum bond order (PAB) for a given pair of atomic species A and B in the molecule.
- Thermodynamic Descriptors: These descriptors are calculated on the basis of the total partition function (Q) of the molecule and its electronic, translational, rotational, and vibrational components.

Altogether over 1400 molecular descriptors were calculated. The actual number of descriptors calculated for each compound depended on the structural complexity of the compound. Descriptors which were not defined for every compound were either omitted or padded with zeros in the subsequent development of QSPR correlation equations.

Multi- Linear Regression Analysis: A multi-linear model was developed to relate the molecular descriptors to the property of interest. The general form of the correlation is:

$$y=\beta_o+\sum_{i=1}^{N}\beta_i x_i \qquad (5)$$

where, y = property of interest, β_o = intercept, N = number of molecular descriptors, β_i = coefficient for descriptor 'i' and x_i = molecular descriptor 'i'. The multi-parameter regression that maximizes the predicting ability is determined using the following strategy,

1. All orthogonal pairs of descriptors i and j are found in a given data set.
2. The property analyzed is treated by using the two-parameter regression with the pairs of descriptors, obtained in step 1.
3. For each descriptor pair obtained in the previous step, a non-collinear descriptor, k is added, and the respective three-parameter regression treatment is performed.
4. For each descriptor set, chosen in the previous step, an additional non-collinear descriptor scale is added, and the respective (n+1) parameter regression treatment is performed
5. The final product of the above steps is a linear relationship between molecular structure and the property of interest containing n parameters.

Neural Network Model Development: The descriptors involved in the best linear model were used to develop the nonlinear models. A variety of neural network models are available to users and the best type of network and its architecture are decided upon through a trial and error process. A back–propagation network proved to be the best network for our study and was used to map the descriptors to the SVRC parameters α_c and α_t. The first step was to determine suitable back propagation network architecture, i.e., the number of hidden neurons and hidden layers. As the number of hidden neurons increases the ability of the network to fit the training data goes up, while, the ability of the neural network to provide accurate predictions decreases. There are no universal guidelines for constructing a neural network; a few rules of thumb exist that provide some guidance in the choice of the architecture. However, most of these rules are approximations and a trial and error remains to be the method for choosing the best descriptors. Several runs with varying network architecture are performed to determine the best network structure. As a rule of thumb, the ratio of molecules to the number of adjustable parameters was maintained to be greater than 2.0 [16].

Discussion

Initial correlations for SVRC parameters were based on data ranging from P_T (pressure at triple point) to P_C (critical pressure). However, the results obtained did not show good correlations with chemical structure. An analysis of the results obtained is provided in Fig. 2, which shows that, when data spanning the entire saturation range were used for correlating vapor pressure, over 30% of the predictions showed average absolute deviations (AAD) greater than 0.5%. The above problem with model predictions can be attributed to the high uncertainty in vapor pressure data below 0.01bar. The use of vapor pressure data below 0.01 bar, in model development skewed the SVRC parameters which made it impossible to correlate them to the chemical structure. Thus, SVRC parameters were regressed using data from 0.01bar to critical pressure. The model predictions obtained using the new set of parameters can be seen in Fig. 3, which shows only 1% of the data with greater than 0.5% AAD. The vapor density correlations also exhibited the above mentioned data dependence. Further, the availability of reliable vapor density data over the rest of the saturation region was limited. No such data dependence was observed on the liquid density correlation model.

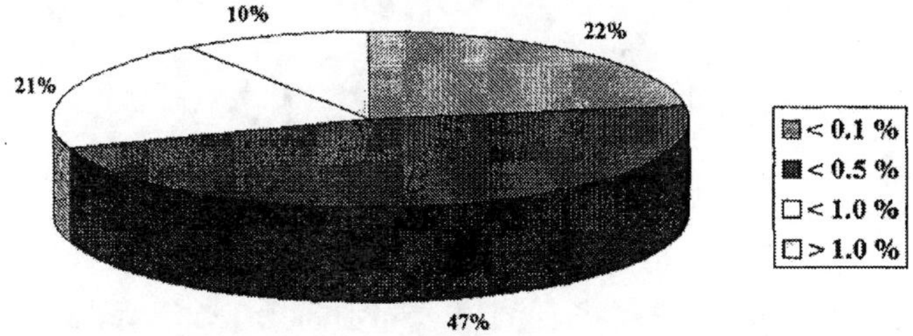

Fig. 2: %AAD Distribution for Vapor Pressure Predictions (Pt to Pc)

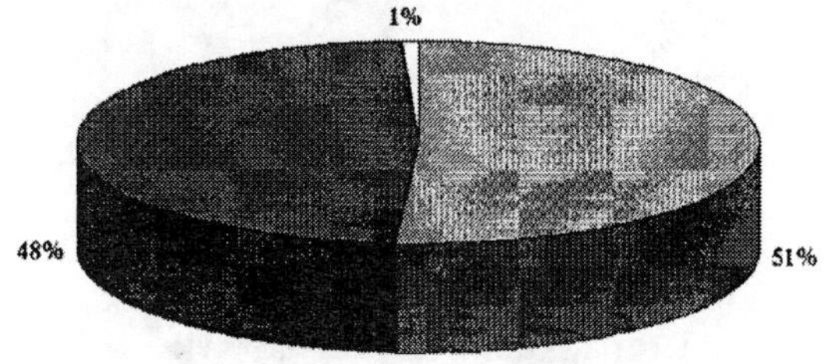

Fig. 3: % AAD Distribution for Vapor Pressure Predictions (0.01bar to Pc)

Ten-parameter linear models were constructed for all the SVRC parameters using a commercial QSPR package. The linear models developed provided satisfactory estimates for the α_c and $\Delta\alpha$ parameters. A typical plot of regressed vs. predicted α_c obtained during linear model development for vapor density modeling is shown in Fig. 4., The linear models were able to estimate the parameters with a %AAD of 0.54 for all saturation properties, with a maximum % deviation of 2.35 for alcohols obtained for the estimation of vapor pressure. These linear models used did not, however, provide satisfactory estimates for the saturation properties themselves. Large prediction errors were noticed in the saturation properties, resulting from small errors in the generalized parameters. A sensitivity analysis on the parameters was performed to determine the effect of errors in SVRC parameter estimates on saturation property prediction. Table 1 tabulates the results of the sensitivity analysis. The results clearly indicate that the model was highly sensitive to errors in α_c. Vapor density model exhibited the most parameter sensitivity with 1% errors in α_c resulting in over 2% error in the property. The liquid density parameter estimates obtained from the linear model

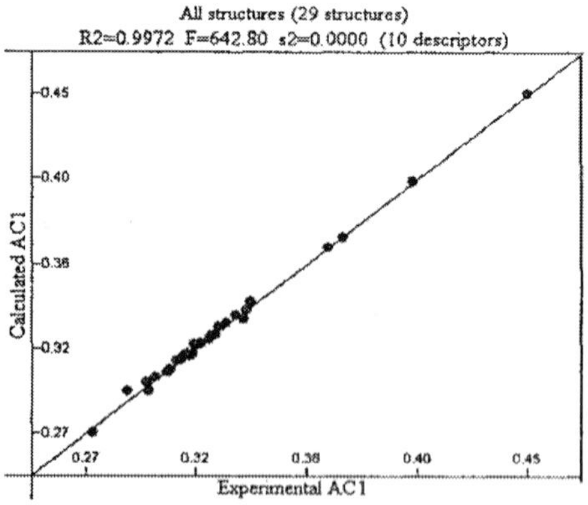

Fig. 4: Linear Correlation for α_c

were satisfactory, while those of the vapor pressure and vapor density needed refinement. A non-linear neural network based model was developed to improve the accuracy of parameter estimation. Non-linear analysis showed that the vapor pressure parameters were easier to correlate and required fewer hidden layers than vapor density. The network architecture was decided by trial and error. A 10-3-1 network architecture provided the best results for correlating the α_c and $\Delta\alpha$ parameters for vapor pressure, while a 10-4-1 network architecture was better for the vapor density correlations.

Table 1: Effect of errors in SVRC parameters on saturation property predictions

Property	SVRC Parameter	% AAD Regressed	0.1% Error in Parameter	1.0% Error in Parameter
Vapor Pressure	α_c	0.05	0.15	1.32
	$\Delta\alpha$		0.07	0.20
Liquid Density	α_c	0.08	0.13	0.18
	$\Delta\alpha$		0.12	0.14
Vapor Density	α_c	0.32	0.45	2.34
	$\Delta\alpha$		0.33	0.56

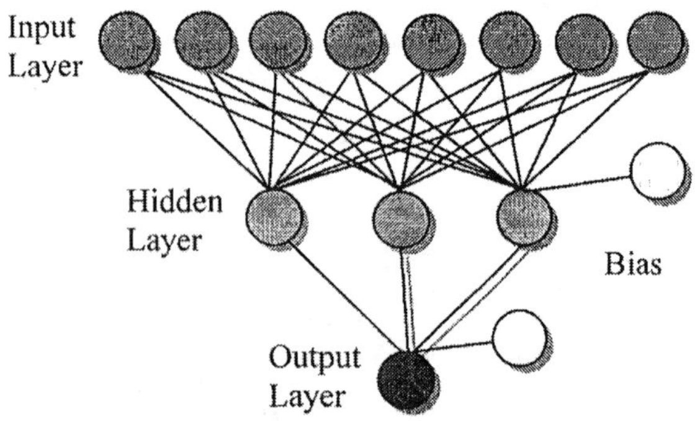

Fig.5: Architecture of a Typical Back Propagation Neural Network

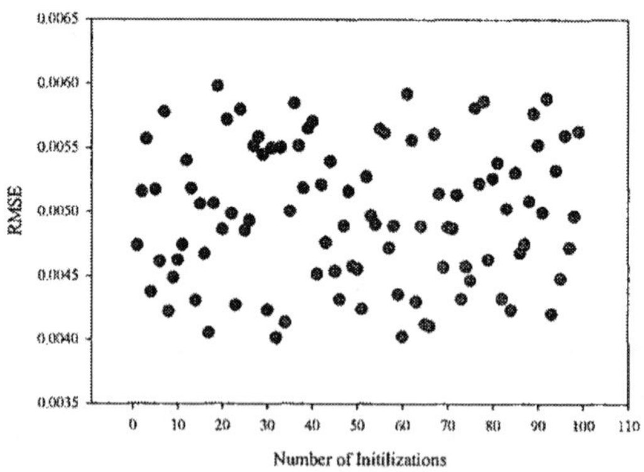

Fig. 6: Results for 100 Network Initializations

The choice of network architecture is still more art than science. A trial-and-error procedure is always used to identify the best set of input and hidden neurons. A typical back-propagation network with eight inputs and three hidden neurons is shown in Fig. 5. Once a suitable the network architecture has been identified the next step is to

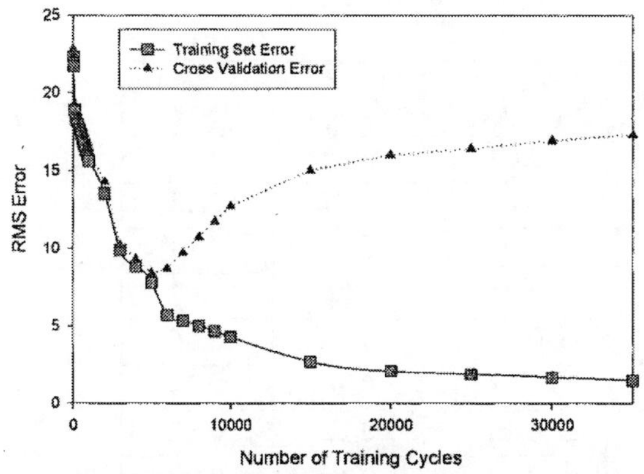

Fig. 7: Error Profiles for Training and Cross validation Sets

determine the adjustable weights on the network connections. The usual procedure is to randomly initialize these weights multiple times and determine which set of weights provides the lowest training error. In our case, the network was initialized using 100 random sets of weights and the results were recorded. Fig. 6 provides the plot of root-mean-square error obtained for each set of weights used. The final problem usually encountered in neural network model development was that of over training. The error for the training set usually decreases as the number of training cycles increases [17]; however, the ability of the network to predict decreases. To identify the ideal location where the training and prediction sets exhibit their lowest errors, a procedure known as cross validation (CV) is commonly employed. The entire data is split into training and cross validation sets. The network is trained using only the training dataset. Periodically the training is halted and error values for the cross validation set are predicted. Fig. 7 shows that the error for the validation set reaches a minimum and then gradually increases. Any training beyond the CV set minimum error the network is being over trained. At the onset of an increase in the CV set RMSE the network has essentially learned all of the general information from the training set and has just started to memorize their individual characteristics. Since the goal of network training is not only to produce small errors in the training data, but also produce reasonable prediction results (referred to as the ability to generalize), the training cycle corresponding to the minimum CV set error corresponds to the place to stop training.

Results

Tables 2, 3 and 4 provide a summary of the modeling results obtained. While non-linear models were used for vapor pressure and vapor density modeling, linear models were sufficient for liquid density modeling. Figs. 8 through 13 provide graphical representations of the results obtained for saturation properties. The figures provide a plot of % deviation vs. the reduced temperature and % deviation vs. saturation property. It can be seen that for all saturation property predictions, the error is maximum near the critical temperature. Some deviations do exist in the vapor density model which was the most challenging saturation property to model. We hypothesize that the orders of magnitude span of the vapor density data is responsible for high prediction errors. Currently work is underway to model the vapor phase compressibility factors instead of the vapor density, which may help us resolve this problem. The results obtained from QSPR analysis provide a clear insight into the structural features that determine the SVRC parameters of an organic molecule. The QSPR models for SVRC parameters involved four critical properties (T_r, T_C, P_C and ω) along with other structural descriptors. The limited availability of space for this article limits us from discussing further the relationship between these structural features and the saturation behavior of molecules.

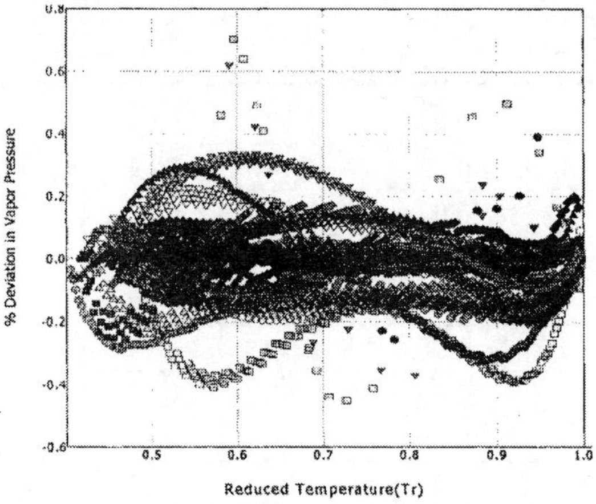

Fig. 8: % Deviation vs. T_r

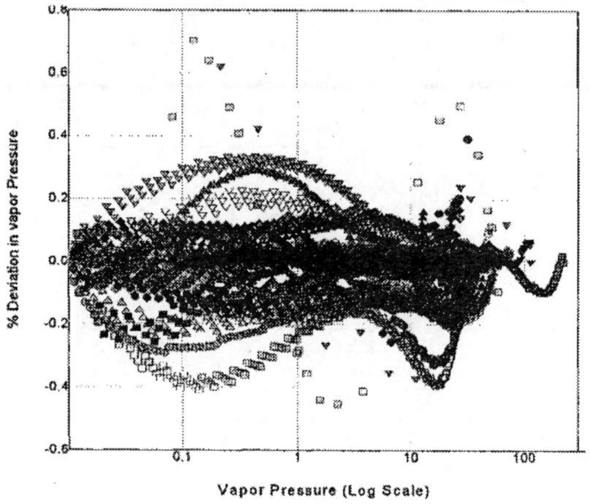

Fig. 9: % Deviation vs. Pressure

Table 2: SVRC-QSPR Modeling Results for Vapor Pressure Data

Molecule	% AAD Regressed	% AAD Linear	% AAD NN	Molecule	% AAD Regressed	% AAD Linear	% AAD NN
Methane	0.02	0.31	0.02	R-125	0.04	0.37	0.08
Ethane	0.02	0.92	0.02	R-124	0.04	0.71	0.22
Ethylene	0.04	0.27	0.07	R-134a	0.05	0.90	0.13
Propane	0.04	0.33	0.08	R-143a	0.06	0.34	0.07
Propylene	0.03	0.41	0.23	R-152a	0.05	0.32	0.06
Butane	0.07	0.42	0.17	Pentane[a]	0.08	0.24	0.16
Iso-Butane	0.10	0.47	0.20	Hexane[a]	0.10	0.22	0.38
O_2	0.05	0.79	0.06	Heptane[a]	0.04	0.12	0.07
F_2	0.06	0.22	0.26	Ethanol[a]	0.28	1.18	0.28
N_2	0.02	0.23	0.07	R-11[a]	0.02	0.44	0.12
NH_3	0.04	0.27	0.08	R-12[a]	0.09	0.63	0.85
H_2O	0.06	0.13	0.14	R-23[a]	0.14	0.41	0.43
CO_2	0.01	0.01	0.01	R-125a[a]	0.14	0.25	0.14
H_2	0.01	0.25	0.06	R-141b[a]	0.04	4.76	0.40
Benzene	0.03	0.26	0.13	R-142b[a]	0.15	0.68	0.57
Acetone	0.06	0.31	0.09	R-143a[a]	0.06	0.34	0.36
Methanol	0.07	0.49	0.08	R-227ea[a]	0.23	0.41	0.23
R-22	0.05	0.92	0.10	R-236ea[a]	0.44	1.33	0.45
R-32	0.04	0.61	0.19	R-245cb[a]	0.23	0.67	0.23
R-123	0.05	0.34	0.14	R-C318[a]	1.14	1.78	1.15

Table 3: SVRC-QSPR Modeling Results for Liquid Density Data

Molecule	% AAD Regressed	% AAD 8-desc Linear	% AAD 10-desc Linear	Molecule	% AAD Regressed	% AAD 8-desc Linear	% AAD 10-desc Linear
Methane	0.04	0.12	0.07	CO_2	0.02	0.05	0.05
Ethane	0.03	0.06	0.21	H_2	0.43	0.43	0.43
Ethylene	0.02	0.21	0.08	R-22	0.03	0.21	0.13
Propane	0.06	0.12	0.09	R-32	0.02	0.04	0.14
Propylene	0.08	0.13	0.10	R-123	0.03	0.06	0.04
Butane	0.04	0.09	0.04	R-125	0.03	0.09	0.03
Iso-Butane	0.09	0.18	0.14	R-124	0.02	0.10	0.04
Pentane	0.07	0.07	0.09	R-134a	0.03	0.05	0.16
Hexane	0.09	0.23	0.27	R-152a	0.03	0.07	0.04
Heptane	0.21	0.28	0.47	R-143a	0.02	0.10	0.08
O_2	0.05	0.08	0.07	Acetic acid	0.18	0.19	0.19
F_2	0.20	0.26	0.26	Benzene	0.35	0.58	0.37
N_2	0.03	0.10	0.10	Ethanol	0.70	0.76	0.80
NH_3	0.05	0.07	0.06	Methanol	0.24	0.25	0.25
H_2O	0.40	0.40	0.40	Acetone	0.05	0.11	0.07

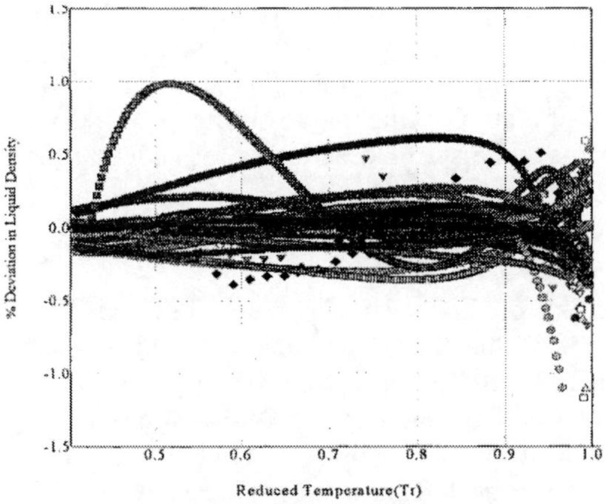

Fig. 10: % Deviation vs. T_r

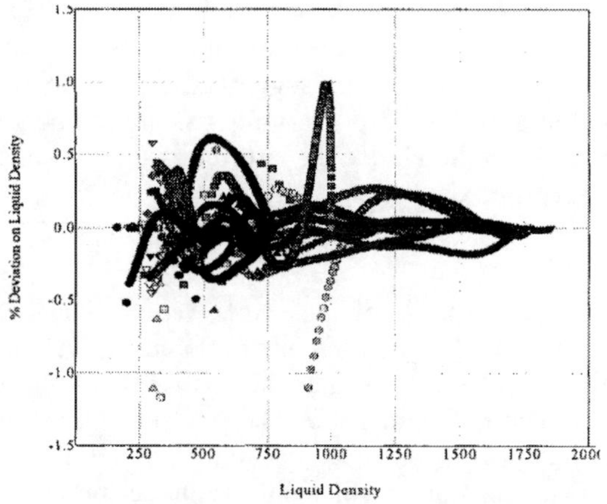

Fig. 11: % Deviation vs. Liquid Density

Table 4: SVRC-QSPR Modeling Results for Vapor Density Data

Molecule	% AAD Regressed	% AAD Linear	% AAD NN	Molecule	% AAD Regressed	% AAD Linear	% AAD NN
Methane	0.07	0.21	0.07	NH_3	0.68	0.68	0.68
Ethane	0.08	0.20	0.08	H_2O	0.72	1.08	0.72
Ethylene	0.20	0.79	0.20	CO_2	0.11	0.12	0.11
Propane	0.04	0.46	0.04	H_2	1.83	1.83	1.83
Propylene	0.82	0.86	0.82	R-22	0.11	1.90	0.12
Butane	0.15	0.42	0.15	R-32	0.17	0.97	0.17
Iso-Butane	0.14	0.21	0.14	R-123	0.29	0.63	0.29
Pentane	0.14	0.98	0.18	R-125	0.15	0.91	0.16
Hexane	0.23	0.39	0.34	R-124	0.07	0.30	0.13
Heptane	0.45	0.48	0.50	R-134a	0.12	1.87	0.15
O_2	0.21	0.77	0.21	R-152a	0.29	0.92	0.29
F_2	0.66	0.68	0.66	R-143a	0.14	1.26	0.14
N_2	0.12	0.20	0.12				

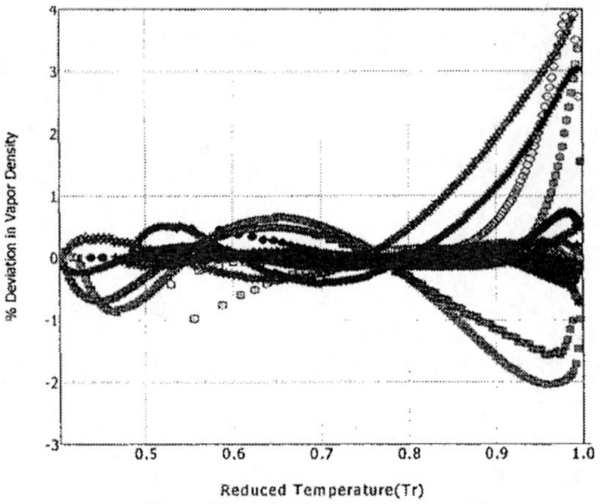

Fig. 12: % Deviation vs. T_r

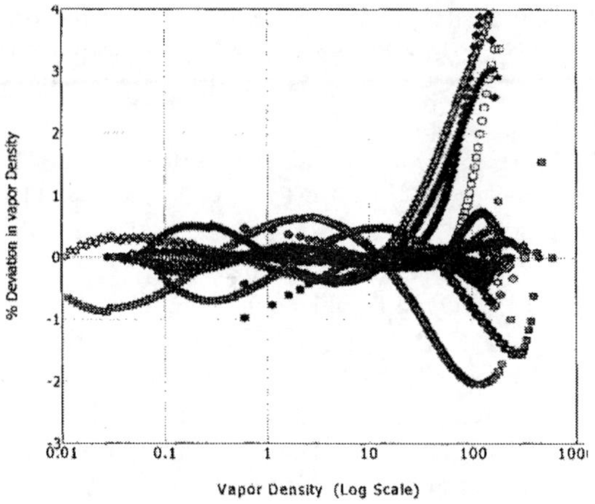

Fig. 11: % Deviation vs. Vapor Density

These QSPR modeling results suggests that a small group of descriptors may be sufficient to model all thermo-physical properties. Recent literature works [18] into modeling of boiling point, critical properties vapor pressure and liquid densities suggests the same. In contrast, several QSPR models exist in the literature which uses different sets of descriptors to model saturation behavior of molecules at a single temperature. In future work, we will investigate the choice of descriptors and the possibility of using the same set of descriptors to model all the saturation properties.

Conclusions

The purposes of this work were mainly (1) to study the utility of molecular descriptors developed by commercial QSPR packages in developing QSAR models for generalizing the SVRC parameters based on their structure and (2) to investigate the possibility of improving upon the existing SVRC saturation property predictions by using such QSPR models. The fundamental assumption was that neither EOS models nor QSPR based models alone could provide satisfactory predictions. While QSPR models have been shown to work effectively for structure based generalizations, their ability to model multiple temperature data is suspect. This is also the case with EOS, which suffers from problems in parameter generalizations. This work puts forth a hybrid strategy where the SVRC model provides the necessary backbone to model multiple temperature data while QSPR models provide the necessary parameter generalizations based on chemical structure. The results obtained clearly indicate that this approach has some merit and is worth further investigation. This work is by no means complete and the models developed still have room for improvements. Several other classes of organic molecules such as acids, aldehydes and ketones need to be added to the dataset. Descriptors which define the structure more accurately will also help to improve the model predictions.

References

1. Hampe, M.J., Selection of Solvents in Liquid-Liquid Extraction According to Physio-Chemical Aspects, Gen. Chem. Eng., 251 (1986).
2. Zanetti, R., Distillation: A Technology in Crisis?, Chem. Eng., 5 (1997).
3. N. S. F. Workshop, Thermodynamic Needs for the Decade Ahead: Theory and Experiment, Washington, D.C., October 28-29, (1983).
4. Shaver, R.D., Robinson, R.L., Jr., Gasem, K.A.M., A Framework for the Prediction of Saturation Properties: Vapor Pressures, Fluid Phase Equilibria, 64, 141-163 (1991).
5. Shaver, R.D., Robinson, R.L., Jr., Gasem, K.A.M., A Framework for the Prediction of Saturation Properties: Liquid Density, Fluid Phase Equilibria, 64, 81-98 (1992).
6. Schult, C.J., Towmey, D.W., Robinson, R.L., Jr., Gasem, K.A.M., Prediction of Vapor Pressures of Refrigerants using the SVRC Correlating Framework, Proceedings of Third International Petroleum Environmental Conference, Albuquerque, September, (1996).
7. Schult C.J., Godavarthy, S.S., Robinson R.L., Jr., Gasem K.A.M., Prediction of Liquid Density of Refrigerants using the SVRC Correlating Framework, Heat Transfer Engineering, 24, 55-63, May, (2003).
8. Katritzky, A.R., Lobanov, V.S., Karelson, M., QSPR: The Correlation and Quantitative Prediction of Chemical and Physical Properties from Structure, Chem. Soc. Rev., 279-287 (1995).
9. Godavarthy, S.S., Gasem, A.K., Robinson, R.L., Jr., and Gasem, K.A.M., Improved Models for Pure Property Predictions Using Quantitative Structure Property Relationships and Artificial Neural Networks, AIChE Spring Annual Meeting, New Orleans (2003).
10. http://dippr.byu.edu.
11. Chemdraw 8.0, Cambridge Software (2004).
12. Stewart, J.P., MOPAC Program Package (1990).
13. AMPAC 6.0, Semichem Inc., 7128 Summit, Shawnee, KS (1998).
14. CODESSA 2.63, Semichem Inc., 7128 Summit, Shawnee, KS (1998).
15. DRAGON 4.0, Milano Chemometrics, Italy, (2003).
16. Long, S. D, Kiers, H.A.L., Chemometrics Intell. Lab. Syst., 14, 155 (1992).
17. Wessel, M. D., Jurs, P. C., Prediction of Normal Boiling Points of Hydrocarbons from Molecular Structure. J. Chem. Inf. Comput. Sci., 35, 68-76 (1995).
18. Godavarthy S.S., Carlson G.D., Robinson R.L., Jr., Gasem K.A.M., QSPR Approach for Melting Point Prediction, AIChE Spring National Meeting, New Orleans, (2002).

Using OPTO22 Programmable Logic Controllers for a Distillation Column to Create and Compare New Control Schemes

Akshay Vilivalam, Charles Nippert, Helen Chu
Widener University

Abstract

Distillation columns can be used as a prototype to demonstrate PID controllers and alternative controls in an undergraduate laboratory setting. Universities use small-scale distillation columns such as the Technovate Distillation Column, Model 9079 in their undergraduate laboratories. This unit was designed for batch mode, manual operation. Such an operating mode does not accurately reflect common industrial practice in many settings. Our paper describes the conversion of Widener's column from its original manual/semi automatic batch mode to automatic batch or continuous operation under control of an OPTO22™ Programmable Logic Controller (PLC). Using OPTO22 hardware and FactoryFloor™ software (a product of OPTO22), our team installed automatic controls so that the SNAP–LCSX™ PLC can operate the column. The PLC's software is developed on a stand alone PC that serves as the user interface using software tools included in FactoryFloor. This expanded capability was much more cost effective than purchasing new equipment

The modifications of the unit allow us to test a variety of control schemes on the unit. The distillation column is currently used in undergraduate laboratories using classical PID control schemes in batch and continuous modes. The apparatus is also being used to explore advanced control methods. Changes to the control setup are easily made using the program development capabilities of FactoryFloor. The current version of FactoryFloor includes a line-based, scripting language that resembles FORTRAN, C, and BASIC in addition to its more traditional block diagram development tool. A strategy is being developed to control the unit using dynamic matrix control (DMC). Examples of operation under these methods of control are discussed. Experimental results are to be compared to simulations to validate a phenomenological process model.

Introduction:

Widener University operates a small-scale distillation column (Technovate Distillation Column, Model 9079) for use in the under-graduate laboratories. The system in the unit is a binary Isopropyl Alcohol (IPA) – Water mixture with IPA having 30% mass weight and water 70% mass weight (11% Mole-Fraction IPA). The column had been operated manually/semi automatically in a batch mode for nearly twenty years. An earlier renovation allowed a programmable logic controller (PLC) to control reflux flow but not reboiler power, bottoms take off or feed, as these controls were unnecessary for batch operation. We wanted to be able to operate this distillation column in a continuous mode using automatic controls controlled by a PLC. We would then be able to use a variety of control methods by reprogramming the PLC. This conversion required the installation of new hardware and software. Additionally, specialized dedicated control strategies for our unit were written using FactoryFloor™ software to drive the PLC. Both the SNAP–LCSX PLC that we used and FactoryFloor™ program are products of OPTO22 (1).

Hardware Additions

The SNAP–LCSX controller is a PLC consisting of a 32-bit Motorola processor, with 1 MB of RAM. The communication between the unit and the computer is through a RS-232 serial cable. The unit can be controlled remotely via Ethernet using FactoryFloor software, but we have not utilized this capability yet. The LCSX connects to two B200 "Brain Boards" that contain Analog to Digital (AD) and Digital to Analog (DA) modules that process signals between the distillation column and the PLC. The AD and DA modules are plugged into the 16 channels on the brain board. A simplified schematic of the PLC hardware configuration is shown in Figure 1

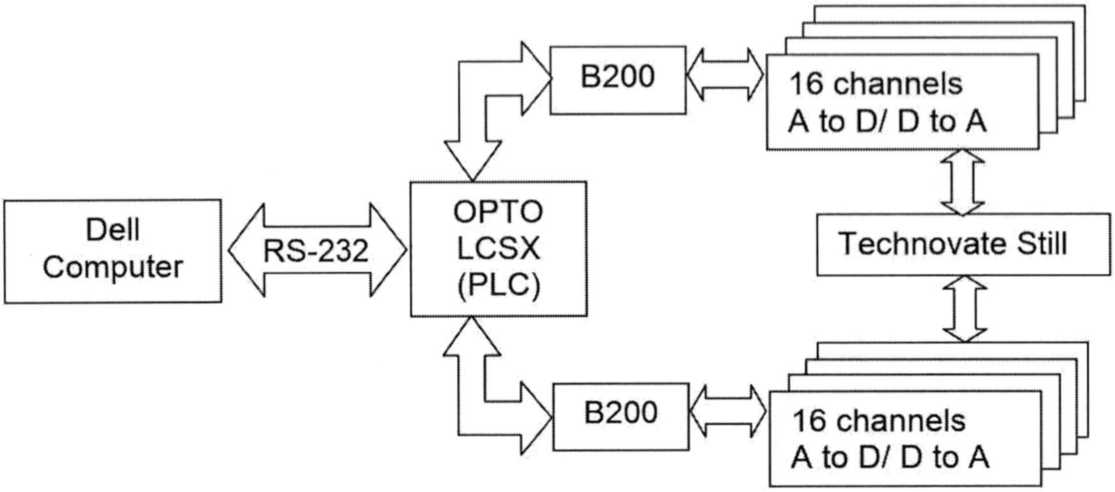

Figure 1 – Schematic of PLC Configuration

The original configuration of the distillation column required manual operation. An earlier modification allowed control of the reflux flow for semiautomatic batch operation but did not allow the PLC to control take off from the bottoms or power supply, which are required for control during continuous operation. The control strategy required an additional 16-channel board to be retrofitted to the PLC. To control the distillation column, the following have been added to provide an interface to the stand-alone personal computer:

- SNAP–LCSX Programmable Logic Controller
- G4B200 16 Channel I/O Brain Board
- OPTO22 Module AD8T Type K Thermocouple monitors
- OPTO22 Module DA3 Digital to Analog Converter (4-20 mA Output)
- OPTO22 Module AD3 Analog to Digital Converter (4-20 mA Input)
- OPTO22 Module AD3T Analog to Digital Converter (4-20 mA Output)
- Bottoms Take-off Pump – Cole-Palmer 7144-00
- Omega Engineering Differential Pressure Sensors
- OPTO22 Solid State Relay, Model 240D10

Software

To control the hardware described in the previous section, OPTO22 supplies a software program called FactoryFloor™. This software is flexible and can be installed on a standard personal computer. The personal computer we use is a Dell OPTIPLEX GX240 with a Pentium III processor at 1.7 GHz and 138 MB RAM. FactoryFloor version 4.0 is used, which has a scripting tool resembling Fortran Basic and C.

We wrote and tested a classical PID control scheme to successfully operate and control the distillation column in continuous operating mode. In the undergraduate laboratory, the distillation column is used in either a batch or a continuous mode of operation. The continuous PID control setup is shown in Figure 2. This setup reflects the use of an accumulator rather than a reflux splitter in the hardware. It should be noted that the large size of the reboiler (14L) relative to the rest of the system effectively decouples the top and bottom control loops because the reboiler's size gives the bottoms control loops significantly larger time constants. This scheme allows the feed rate to remain as an independent variable such that the controller responses to process upsets may be more easily explored.

The specific control objective is to operate this distillation column in continuous mode to yield a mixture of IPA and water with a composition near the azeotrope as product. The initial pot concentration, and the feed concentration is 11% mole-fraction of IPA and the reserve feed tank also contains the same composition of IPA. When continuous operation is initiated, the pot concentration is approximately 7% mass weight (.03% mole-fraction). To ensure optimal usage of energy and to optimize the product take-off, the accumulator level was used as a set-point, to allow constant product flow. The temperature of the top tray was used as a control variable. The manipulated variables were product take off and reflux ratio. Feed rate in continuous operation was intentionally left as an independent variable.

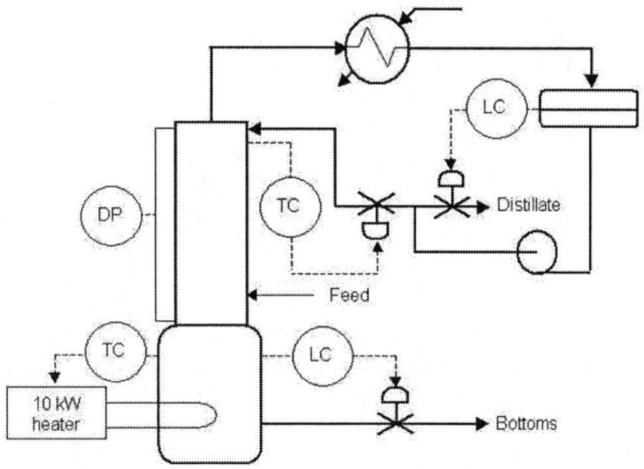

Figure 2 – Control Scheme Continuous Operation

OPTO Script Programming

FACTORYFLOOR™ software consists of three major applications, OptoControl (OC), OptoDisplayConfigurator (ODC), and OptoDisplay (OD). OC is the part of the software where the actual development of the control scheme is performed. The specific control setup (called strategy) is then uploaded via an RS-232 serial cable connected to the SNAP–LCSX system, which is, in turn, connected to the B200 Brain Board. All analog to digital (AD) and digital to analog (DA) interface modules are mounted on B200 boards. The OC software gives us the ability to write any control setup as different subroutines, called as sheets. Software sheets also provide threading capability that allows scanning of inputs, the computer and messaging to the control elements. Software variables declared in the OC are globally available in the strategy; any sheet can use any variable as its values changes. The brain board hardware has built in PID controllers. In our case these are overridden in order to provide more versatile control (Figure 3). The OPTO SCRIPT allows rapid development of novel control configurations and quick modifications of the control scheme using conventional programming techniques. This programming approach is different than the ladder logic used in many other PLC environments. Since the workers are already familiar with line - based programming, this presented a considerable advantage. It should also be noted that, in the authors' opinion, line based programming offers superior power when developing new control methods.

The ODC is the part of the software where the interface, OD, is developed (Figure 4). The final result is called as the runtime display, which is the actual user interface used by students to operate and control the distillation column. ODC allows the development of the user interface to be extremely interactive, allowing either masked variable changes or actual discrete values such as mA signal values to be sent directly though OD to the LCSX to control the distillation column, once the related strategy is loaded. The interface allows variable changes and provides us with real-time graph of necessary values such as temperatures and accumulator level over time. The values for the gain, integral and derivative actions can and are varied on the OD to tune the PID. Once the PID has been tuned, these values can be stored through the OD to be the default values.

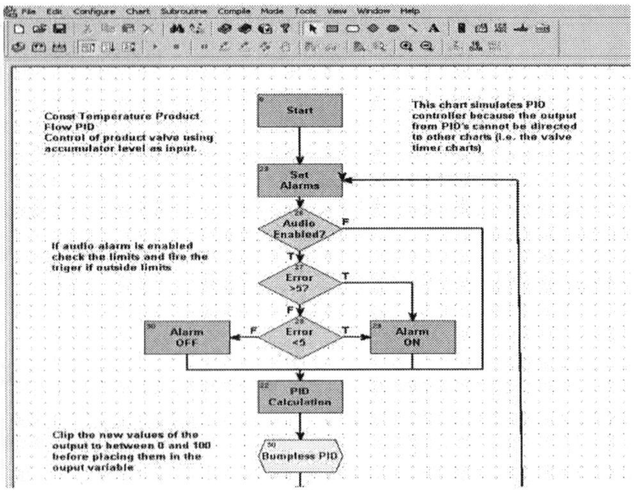

Figure 3 – OptoControl Software

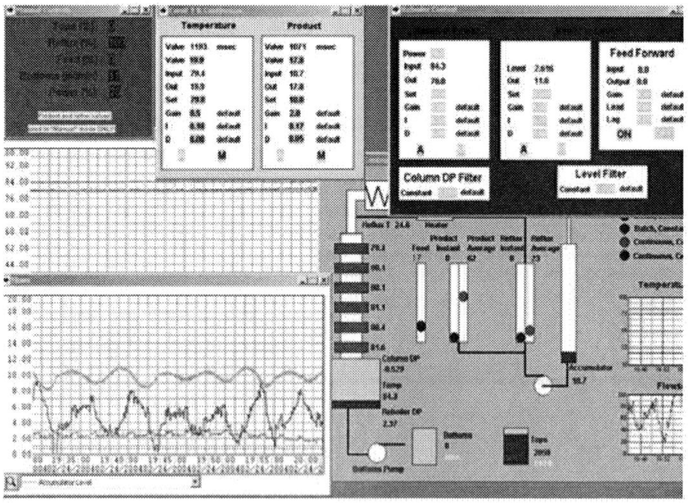

Figure 4 – OptoDisplay Software

All variables from the sensor inputs are displayed as-is such as temperature, or converted to appropriate masks via dedicated routines. For the valve openings, the masked variables are also sensor inputs, and as all signals are in the 4-20 mA range, 0% is the mask for a 4mA signal and 100% is the mask for a 20mA signal. The feed pump, the product take-off pump, the reflux pump, and the bottoms pump were all calibrated for a 4-20 mA signal with 4 being no flow and 20 being the maximum flow rate. This was done because the pumps responses are non-linear.

Operation and PID Tuning

Development of a process model of an unsteady state non-linear distillation column is extremely difficult to derive accurately. Compositions of the feed, variations in liquid hold-up in the accumulator due to variations in vapor flow-rates and the varying rates of change of the parameters in general in the process dynamics of distillation are some aspects to be considered if modeled as non-linear and unsteady. To facilitate model development, at any given instant of time, the entire system is modeled as first order plus dead time (FOPDT). This allows us to treat the system as a linear system, without deviating much from operating conditions. All PID's are tuned using FOPDT model (Figure 5).

$$\frac{Y(s)}{Q(s)} = \frac{Kp}{Ts+1}e^{-Ls}$$

Figure 5 – FOPDT Model Equations

From experimental results, it is noted that assuming the system to be a FOPDT adequately represents the entire system. The valves are timed on a six second cycle and hence the PID is modeled to run on a six second cycle. The reboiler is controlled using the relay and the power supplied to the relay is a PID loop using accumulator level as input. Also, when run in continuous, the feed temperature is less than the pot temperature, and so temperature of the reboiler is used as a set point. The product take-off is maintained constant by a PID, which uses accumulator level as an input, and by varying reflux ratio.

Dynamic Matrix Control (DMC)

DMC is a type of Model Predictive Control, which is found to be an excellent application in multivariable control. Any system that can be represented by a system of linear equations can utilize the DMC approach. DMC modeling is done in either of two ways, as a fit to the FOPDT process or as a step response to a unit change in the operating conditions. We will be using the step response model, which utilizes the principle of superposition to predict the next step with respect to the previous knowledge of the initial step change in input. Since DMC is a model response based control scheme, any changes to the current model will affect the control scheme and the control scheme cannot be used to control other distillation columns. As in the PID control scheme, accumulator level will be used as a set point, and the reboiler will be controlled using the accumulator level as an input and the power as the manipulated variable. The product-takeoff will be controlled using the accumulator level as a set point and the reflux ratio as the manipulated variable. The operation and the experimental results of the DMC control configuration are not described in this paper; the DMC control scheme is currently under development and is an extension of the work discussed in this paper.

Distillation Computer Model, Preliminary Comparison

A computer model, which is an exact replicate of the experimental model, was developed in JAVA^{TM}. The basic derivation of the dynamic simulation model was described elsewhere (ASEE paper) and the modifications of the model to continuous operation are the subject of another work to be written shortly. In essence, the model uses assumptions similar to those in McCabe-Thiele calculation to reduce the number of differential equation, so that the simulation can run in real time. This model was used to provide estimates of dynamic properties of the still for during the initial shakedown. As an example, a dynamic simulation of the startup of the still is shown in Figure 6 and can be compared to actual data in Figure 7. related differential pressure changes are shown in Figure 8.

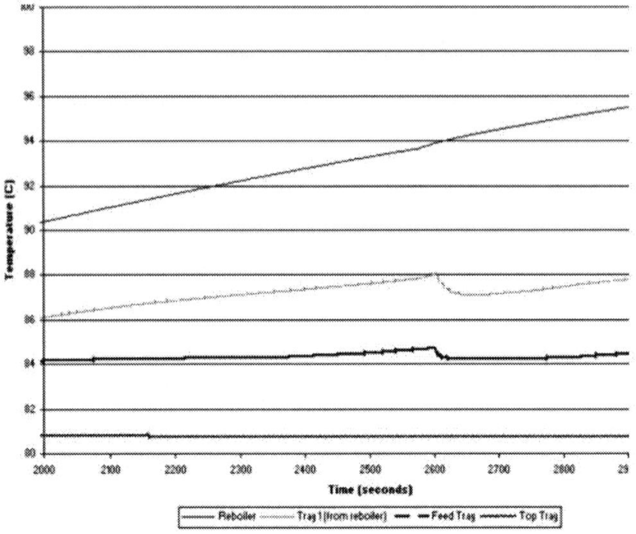

Figure 6 – Simulation Results (T Vs Time)

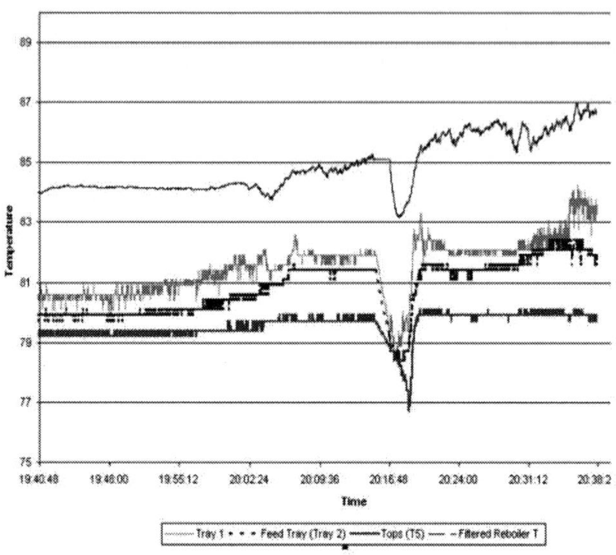

Figure 7 – Experimental Results (T Vs Time)

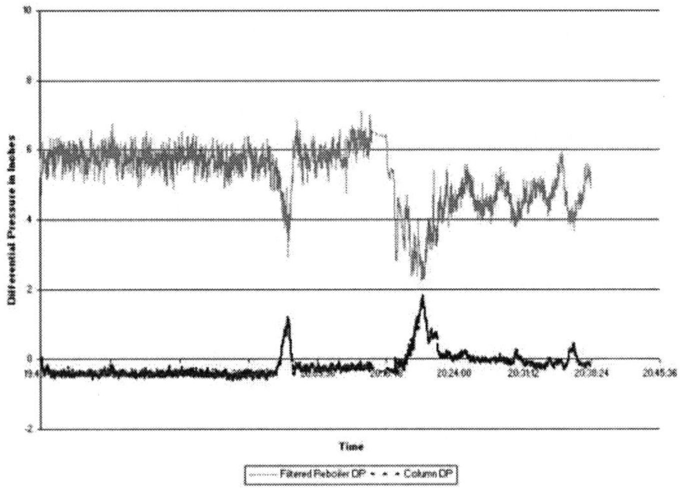

Figure 8 – Experimental Differential Pressure Changes

Experimental Results

Table 1 shows a plot of reboiler temperature compared with the experimental temperatures at steady state. The rise in temperature is associated with the change in composition of the reboiler as its composition approaches the steady state value of 87 degrees Celsius. In this case, the results indicate that the column was approximately at steady state.

Table - 1

Tray Number	McCabe Thiele Temperature	Experimental Temperatures
1	87	86.8
2	84.8	84.1
3	83.1	82.1
4	82	81.9
5	81	80.6
6	80.8	80.6
7	80.3	80.1

Conclusions

These modifications to the distillation column enhance and upgrade the capabilities of our undergraduate laboratory experience by providing a wider range of interesting experiences for our chemical engineering students while acquainting them with new types of control. This modification of existing equipment was performed with minimal cost and allows the study of alternative control schemes in a realistic environment.

References

(1) HTTP://WWW.OPTO22.COM

(2) Nippert, C.R., "A Real Time, Dynamic, Binary Distillation Simulation for WEB Based Instruction", *Proceedings of the 2001 American Society of Engineering Education Annual Conference and Exposition, Albuquerque, NM.*

(3) Marlin, T., "Process Control – Designing Processes and Control Systems for Dynamic Performance", *2nd edition, 2000.*

(4) Luecke, R.H., Lewis, J.C., Lin, H.Y., Yoon, W.K., " Dynamic Matrix Control of a Batch Distillation Coulmn", *Proceedings of the American Control Conference, 1985, p 209-213*

(5) Rice,R., Cooper, D.J., "Design and Tuning of PID Controllers for Integrating (Non-Self Regulating) Processes", *Proc. ISA 2002 Annual Meeting, 424, P057, Chicago, IL (2002)*

(6) Luyben,W.L., "Effect of Derivative Algorithm and Tuning Selection on the PID Control of Deadtime Processes", *Ind. Eng. Chem. Research 40 (2001) 3605-3611.*

Precise Distillation Tower Simulations: Crude Vacuum Towers

Reza Yahyaabadi, Esfahan Oil Refining Company, Esfahan, Iran

Abstract

Distillation towers simulation is an effective tool to find how they work, what is happening inside the towers, are there any upsets inside them, which sections or parts are working well and which not, what are the operating conditions and products yields and quality in different cases and so on. In order to correct evaluation by the simulations, a precisely distillation tower simulation model that have been matched the plant data, is required. By considering vapor-liquid contact in all sections of distillation towers and considering their effects in the model and also entering some non-idealities, a precise distillation tower model could be made. Distillation towers consist of different sections, parts and equipment that many of them affect on the mass and heat transfer and consequently if they are not considered, it will lead to a poor simulation model. By going through on each sections and equipment in detail and using the mentioned method (considering some non-idealities) an accurate model that have met the plant data, could be made. In this article, some of them are explained and the way how they can be entered in a simulation model is brought.

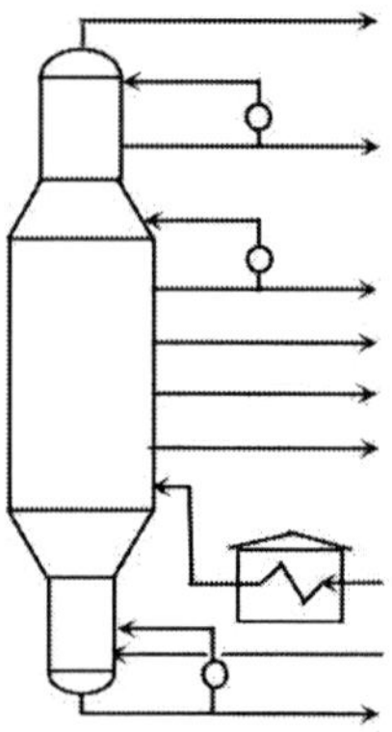

There are wide variety of distillation towers in refineries, petrochemicals and chemical plants. This variety ranges from simple towers that separate two component with one feed, condenser at the top and reboiler in the bottom (like as de-polarizer) up to complex towers that separate some cuts with two or more feeds and equipped with top and mid-pumparounds, some duties in different section and side strippers for the cuts like as crude distillation towers in the refineries.

Nowadays, there are a lot of simulation packages that can simulate distillation towers from simple towers to complex one. But the key point is that how are they used? In some articles the key points and guidelines for making an accurate model to matching the plant data were described [1], [2], [3], [5].

The simulation package outputs are results of calculations based on the input and therefore would not necessarily match the tower operating conditions and laboratory tests results. Many reports described simulations that did not reflect what a tower was actually doing [1].There are many items in a distillation tower simulations that each of them affect the results and consequently the conclusions. So, one who wants to simulate a distillation tower should have enough background on:

- Tower operating conditions variations
- Factors that affect on the results
- Importance of each factors
- Operating conditions consistency
- Simulation model
- Simulation program capability
- Simulation program convergence technique

Due to the above items, making a precise simulation model requires not only process engineering experience, but also simulation program experience and information. The importance of such experiences is apparently clear. Often, non-ideal equipment performance can be modeled if the user understands equipment basics and the model's underlying assumptions [6]. Also some companies have announced that they have significant experience in developing realistic simulation models using proven techniques and assumption and in validation of computer simulation work against actual operating data[4].

Requirements for Simulation

The necessities for simulation of distillation towers are explained in the articles [1], [2], [3], [4], [5]. Some usage of distillation towers simulations are:

1. New distillation column design [4]
2. Revamp of existing column [4]
3. Evaluate the existing column for:
   3.1. Calculating the important parameters (such as wash oil rate in crude vacuum distillation towers) in order to maximizing the yield and avoiding the operating problems like as coking.
   3.2. Predict the operating conditions and products yields and quality at different cases and feeds.
   3.3. Optimizing the operating conditions.

Another option for calculating the required parameters in the above cases is running the pilot plants that can be used in this manner. The costs for this option are apparently high and also more time is required. But the pilot plant results could be more close to the real tower if it is designed and constructed properly. Since the economic profits of making the simulation model are very high, they are always selected for the mentioned cases.

Plant Data

A typical crude vacuum unit arrangement is shown in Figure (1). A list of typical unit data (operating conditions) is brought in Table (1). The streams specifications from laboratory tests are supplemental of unit data (operating conditions). A typical laboratory results on streams are listed in Table (2). Both unit data (operating conditions) plus laboratory results are plant data parts.

Correct plant data must be gathered first. Successful unit revamps and effective troubleshooting start with field data, not theoretical calculations [7]. Real equipment performance should be the basis of a revamp, not office-based assumptions or cursory reviews of the original equipment manufacturer's data sheets [5]. The plant data are always gathered in a test run. Based on the means of making the simulation model, the operating conditions for the test run is selected. The selected operating conditions in the revamp of existing tower for higher capacity may differ from the actual conditions evaluations for optimization or setting the important operating parameters. In some cases, more than one case may be required and test runs for different cases are doing. Among the instrumentations errors in the unit data plus personal and equipment errors in the laboratory, the major problem is that to obtaining a reliable, consistent set of plant data.

After gathering the plant data, they must be interpreted. Applying data to the model includes analysis of data inconsistency [2]. Many distillation tower parameters are related to and affected by each other. So before making the method, the test run data should be analyzed and compared with its normal variation range. For analyzing the test run data inconsistency, experience based engineering is required. In order to making an accurate model, all operating conditions and many laboratory tests results are key points. For this reason, inconsistency analyzing step is important. The experienced engineers based on their knowledge on the process, the previous data in different operating conditions, and their calculations and notes can rapidly specify the inconsistent points and necessity for doing another test run.

Why Precise Simulation

What does mean "precise simulation"? For an existing column modeling, it means the simulation results should meet the plant data. In the articles [(2), (3), (4), (5)], it is mentioned that the simulation outputs must correctly present the reality of columns. For this reason, the program outputs should necessarily be as close to plant data as possible. Revamps often failed due to inaccurate simulation [(1)]. Without a credible, validated base case simulation, the economics of the revamp can not be properly assessed and achievement of revamp goals is placed at risk [(3)]. Some refinery vacuum towers have had problems regarding to coking, high-pressure drop, loss in products quality and short run length.

They resulted from simulations that underestimated the fraction of wash oil vaporized [(1)]. Inaccurate or incorrect results from simulation models are caused by:

1. Inaccurate feed characterization [(1), (3), (6)]
2. Incorrect thermodynamic or physical properties [(1), (2)]
3. Simulation model [(1), (3), (5), (6)]

Real feed characterization detection is one of the most important stages for making accurate simulation models. Characterizing the heavy components in simulating refinery vacuum towers is a key problem [(1)]. There are some standard laboratory techniques for determining the crude vacuum distillation feed characterization. The usual standard refinery distillation test on the vacuum column products and feed is an ASTM D1160. As it is seen in Table (2), due to limitation on temperature of any of the laboratory distillation analyses, for the feed and also heavy products, the whole stream could not be distilled and therefore characterized. The modern laboratory analytical techniques of simulated distillation by gas chromatography produces high-temperature distillation data; however, these techniques diligence in laboratory quality control [(3)]. Historically, vacuum tower feed distillation curves have been generated via an ASTM D-2892 TBP curve procedure or blending the product streams characterized by standard refinery analytical laboratory tests of ASTM D86 and D1160. But some refiners have begun to use simulated distillation [(3)].

After the feed characterization detection and based on the operating conditions, correct thermodynamic method for accurate vapor liquid equilibrium prediction should be selected. Physical properties have also considerable effect on the results. The simulation programs have always guidelines for offering the thermodynamic and physical properties methods that can be used for the specified conditions. Sometimes, using some other equations or methods that differ from some parts of program guidelines would be more suitable. So, the engineer who wants to reach an accurate model should knows them and apply in the model.

By simple entering the operating conditions according to the simulation program package procedures and converging the computer program, one inexperienced engineer may optimistically think that he could use the output of the simulation of a crude distillation tower for revamp easily for example. However, making a precise computer simulation model is far farther than it. As it was also mentioned before, for making a precise computer simulation model, engineering judgments and experiences with going through the tower process in detail by data collections, calculations, reading the papers, doing some pressure and temperature survey, reading and analyzing the previous data, etc. are required especially for the crude vacuum distillation towers.

Computer models must match the observed plant data, otherwise the simulation is at best wrong [(1)]. Matching the unit data by computer models are necessary (especially tower top, flash zone, bottom pressure and also top and F.Z. temperature). But approach to the laboratory results only in their repeatability range is required. Of course the IBP and FBP of streams should not be involved in plant data and simulation output comparison.

In the following, some special consideration are brought so that using them help engineers to making accurate distillation tower simulation models.

Making Precise Distillation Towers Simulation Models

Since the equipment and the process do not exist separately and equipment details count as much as process details [7], all equipment that affect on the process should be considered and their effect should be involved in the simulation model. It would be more important when a crude vacuum distillation tower is being modeled.

For getting the mentioned goal (matching the plant data by computer models), vacuum heater, transfer line and vacuum tower should be involved simultaneously. In a literature [5], influence of transfer line phase separation on predicted heater outlet temperature and calculated wash section liquid flow rates was mentioned. There are many important parameters in operation and design of vacuum heater and its transfer line that are mentioned in the articles [6], [8]. But in making an accurate model, the phase separations as well as the enthalpy changing and pressure drops through them should be considered. The transfer line outlet phases with their temperature are important for matching some plant data. So, the engineers should thoroughly notice to the processes that are taking place in them in detail to enter them in the model correctly.

Besides the vacuum heater and transfer line, there are important parts and sections inside the column that for an accurate model all of them should be considered. They are explained separately. Their effects on the process and how these effects could be involved in the model are also described as followed:

Packing Beds

In order to make a model, the packing equivalent height should be changed to number of transfer stages. It is necessary to use packing efficiency curves for this aim. The packing efficiency curve is already prepared by the manufacturers. A typical curve is shown in Figure (2). For a practical case, suppose in a 2 meter bed height, number of theoretical stage per meter that is calculated from such curve, is equal to 2.7. It means this bed is equivalent to 2*2.7=5.4 theoretical stages. When in the simulation programs theoretical stages are numbered with integer figures, how could the 5.4 stages is entered in to the model? It is apparently clear that using 5 or 6 stages for this bed will produce the different outputs. Exact number of theoretical stages could be applied to the simulation model by using the overall efficiency. For this case, six theoretical stages with the overall efficiency of (5.4/6)*100=90% may be used.

Chimney Trays

As it has been shown in Figure (1), the main function of chimney trays are:

1- Collection of the liquid outlet from the packing beds for either drawing off or directing to the lower liquid distributor that is at the top of the lower bed or both. Also, in the chimney trays proper residence times for degassing the collected liquid are normally considered.
2- Distribution of vapor that is entering to the packing bed.

The flow of liquid and vapor through a chimney tray is shown in Figure (3). As it is seen, in a chimney tray, other event takes place in company with the two mentioned function. The vapor and liquid are in contact with each other. The contact between vapor and liquid in two spaces are there:

1- When vapor passes from the area between the chimney and its hat, it should pass through the liquid that is falling from the hat or the above bed. The liquid comes out the bed; it falls down either directly on the sump or chimney's hat. Both of them meet the vapor that is going out from the area between the chimney and its hat.
2- Between the chimneys hats and the above bed, there is a space that bed's inlet vapor and liquid outlet are in contact with which other.

In other words, besides the liquid gathering and vapor distribution, an amount of heat and mass transfer are taking place. The heat and mass transfer rate depend on chimney tray arrangement that how much does the liquid and vapor contact with each other? So, for an accurate simulation model, some heat and mass transfer should be considered in the chimney trays. How could the effect of contact between vapor and liquid in a chimney tray be considered in the simulation model? Here also one stage with a specified overall efficiency may be considered. Based on the chimney tray arrangement, the overall efficiency for a chimney tray may range from 10% or less to 70% or higher.

So, for the previous example that there were six stages in the packing bed with an overall efficiency of 90%, if the chimney tray overall efficiency is equal to 30%, there will be two options for considering the bed and its chimney tray in the model:

1- Enter seven stages that from top, the first six stages have 90% overall efficiency and the last one has 30% efficiency (total theoretical stages are equal to: 6*0.9+0.3=5.7).
2- Enter six stages with overall efficiency of 100*(5.4+0.3)/6=95%.

Although, the theoretical stages for both cases are identical (5.7), but according to multistage distillation concept in the chemical engineering text, the first one should be used in the simulation models. In the first option, the temperature of liquid that is collected in the chimney tray will be closer to the plant data than the second one.

Spray Nozzle Distributors

In design of distillation tower and especially in crude vacuum towers, the spray nozzle networks are applied for distribution of refluxes or wash oil. A typical spray nozzle arrangement is shown in Figure (4). When the spray nozzles outlet liquid falls down on the packing bed, it is being contacted with the vapor that is coming up from the bed. It means, in the space between the spray nozzle network and the bed, some heat and mass are transferred between vapor and liquid. In other words, in an accurate model, one stage should be considered for this section of tower. The overall efficiency for this stage depends on the arrangement, vapor and liquid properties.

In some liquid gravity distributors, the liquid and vapor can contact with each other, but here the amount of heat and mass transfer between two phases are too small that could be neglected.

Flash Zone Entry and Configuration

There are several designs for flash zone that some of these are based on centre entry of the transfer line and some are based on nominal tangential entry [8]. Typical flash zone arrangement is shown in Figure (5). In the feed inlet device, the two phase mixtures from transfer line are separated. Actually phase separation in F.I.D. is not perfect and therefore some liquid goes up with vapor phase that is named liquid entrainment from the flash zone. The flash zone vapor outlet consists of strip out from stripping section and vapor from feed flashing in the flash zone. The liquid from the feed inlet device falls down to stripping section. Here also vapor and liquid are contacted with each other. Based on the feed inlet device type (centre entry or tangential) and arrangement, the amount of contact between two phases (liquid from feed inlet device and strip out) may differ. It is apparently seen the flash zone temperature (gas phase temperature above the feed inlet device) is affected by the amount of liquid, vapor contact (and therefore heat and mass transfer between two phases) in the flash zone. In other words, in the model, a stage should be considered for the flash zone that necessarily does not act as a theoretical stage. So, based on the type of feed entry, an overall efficiency for the flash zone stage should be considered. Based on the general arrangement, it seems that in centre entry, the flash zone stage overall efficiency is higher than the tangential type.

Stripping Section

Vacuum column stripping section consists of a number of trays. This section is considerably affecting on the simulation models results. This section of tower is facing to heavy components that its process simulation requires special considerations. First of all in the section of feed characterization, proper laboratory tests should be recognized in order to correctly characterize heavy material. Second, proper thermodynamics properties correlation for heavy component should be selected. Third, in simulation model for this section also, by using overall efficiency, the required number of theoretical stages could be created that is not necessarily integer. For an existing tower simulation, by changing the stripping trays overall efficiency, the plant data (including unit operating conditions and laboratory test results) could be matched. This will give valuable information about the actual performance of the installed trays under the condition of vacuum tower stripping section that is already more sever than the other sections (due to the presence of heavy components).

Using the Method

So, for matching the plant data, the engineers who want to simulate the distillation tower has an effective tool for making a "Precise Model".

When the plant data is available (existing tower simulation), changing of the chimney trays efficiency, the side draw-off temperature and specifications (distillation curves) could simultaneously be matched (for consistent data).

For the flash zone section, entering the overall efficiency to its stage depends on type of vacuum tower. In an article [(3)], some guidelines for dry vacuum towers simulations have been brought. And it was mentioned that by assuming an ideal stage for flash zone, the model results closely match observed deep cut vacuum tower operation. In the other article [(6)], for wet vacuum towers it was explained that in reality, the flash zone is not an equilibrium stage.

The author has no experience on the dry vacuum towers. But according to the above explanations, in the wet vacuum towers the assumed stage for the flash zone is not an ideal one and with varying the flash zone stage overall efficiency, both flash zone and slops wax chimney tray temperatures could match (with considering liquid entrainment from the flash zone of course). The effect of non-theoretical stage for the flash zone on the dry-out rate and also the required wash oil is also explained in the article [(6)].

Since, matching the plant data in the simulation model is emphasized, the engineers who are simulating the unit should notice all factors in the unit that are hidden from some others. One of them is heat loss through the pipes and walls. When a vacuum distillation tower is being simulated without considering any heat losses from the tower, some differences in refluxes return temperatures may be observed. The amount of differences depends on the amount of heat losses that is a function of ambient temperature and ambient conditions, insulation material and thickness and etc. If the differences are noticeable, the amount of heat losses that includes heat loss from tower wall and its attachments should be calculated. Due to the temperature gradient though the tower, the mentioned heat loss calculation requires surfaces temperature measurement.

For new tower design the story is different. In this case, the designers do not have the plant data. And they are faced to different conditions for example different feed characteristics, product specifications and so on. What should they do? As it was mentioned before, they should have a powerful background of such cases. For these case studies, they could make assumptions for their design. Otherwise, by doing exact calculations in all sections of tower (not only the packing beds) they have to reach to the calculation based figures.

By using the above methods, a computer simulation model by HYSYS have been tested for a wet vacuum tower. This model has been found to match the plant data very well. It means both unit data (in Table (1)) and laboratory data (in Table (2)) have been matched simultaneously in the range of laboratory tests repeatability.

In this article, some possibilities contact between vapor and liquid and their effect on the heat and mass transfer are explained. In some other designs, there would be other possibilities for vapor liquid contact and therefore heat and mass transfer that their effect should be involved in the model by considering some non-idealities.

Conclusions

Using overall efficiency in the simulation packages which can handle it, will give an effective tool that helps engineers to make an accurate model. In this light, all parts and sections in distillation towers should be considered and their effect on heat and mass transfer between phases should also be entered in the simulation model. In the revamp cases, engineers have the plant data that are very useful for making such model. In other words, the simulation model should necessarily match nearly all plant data. In design cases, engineers should either have the reference data or calculate the effect of all possibility of heat and mass transfer on the model.

Literature Cited

1) Kister ," Can We Believe the Simulation Results ?", Chemical Engineering Progress , October 2002
2) Sloley , Fraser ," Consider Modeling Tools to Revamp Existing Process Units", Hydrocarbon Processing , June 2000
3) Golden, Villalante , Martine, "Feed Characterization and Deepcut Vacuum Columns : Simulation and Design ", Presented at the AIChE 1994 Spring National Meeting April , 1994
4) Schulte , Goelzer, Subramaniam , "Design and Revamp Vacuum Units For Maximum Return " , Hydrocarbon Engineering, Feb. 2000
5) Martine , "vacuum Unit Design Effect on Operating Variables " , ptq, Summer 2002
6) Golden, " Troubleshooting Vacuum Unit Revamps " , ptq , Summer 1998
7) Sloley , " Troubleshooting Petroleum Refinery Vacuum Towers " , Presented at the Chemical Engineering Expo 2000 , 7-8 June 2000
8) Nangia , Schulte , Goelzer , Subramaniam , " Design and Revamp Vacuum Units for Maximum Return ", Hydrocarbon Engineering, April , 2000

The Author

Reza Yahyaabadi is a senior process engineer for Esfahan Oil Refining Company (E.O.R.C.), Iran. He has 15 years experience in process engineering, process revamps and debottle-necking and has a BS degree in chemical engineering from Esfahan University of Technology.

Table (1) : Crude Vacuum Distillation Typical Unit Data	
Item	**Unit Data**
Vacuum Heater	Outlet Temperature and Pressure , Rate
Transfer Line	Outlet Temperature , Pressure Drop
N.C.G.	Rate
Light Slops	Rate
Diesel / VGO Lube#1 / Lube#2 Slops Wax	Draw-Off Temperature , Rates
Vacuum Residue	Rate
Top & Mid Pumparound	Return Temperature , Rate
Column	Top , Flash Zone , Bottom Pressure and Temperature
Wash Oil	Rate
Stripping Steam	Temperature , Rate

Table (2) : Typical Laboratory Tests on Crude Vacuum Distillation Unit Streams	
Item	**Lab. Data**
Atmospheric Residue	Sp. Gr. and ASTM Distillation Up to 70%
N.C.G.	Gas Chromatography
Light Slops	Sp. Gr. and Full ASTM Distillation
Diesel / VGO Lube#1 / Lube#2	Sp. Gr. and Full ASTM Distillation
Slops Wax	Sp. Gr. and ASTM Distillation Up to 30%
Vacuum Residue	Sp. Gr. and ASTM Distillation Up to 30%

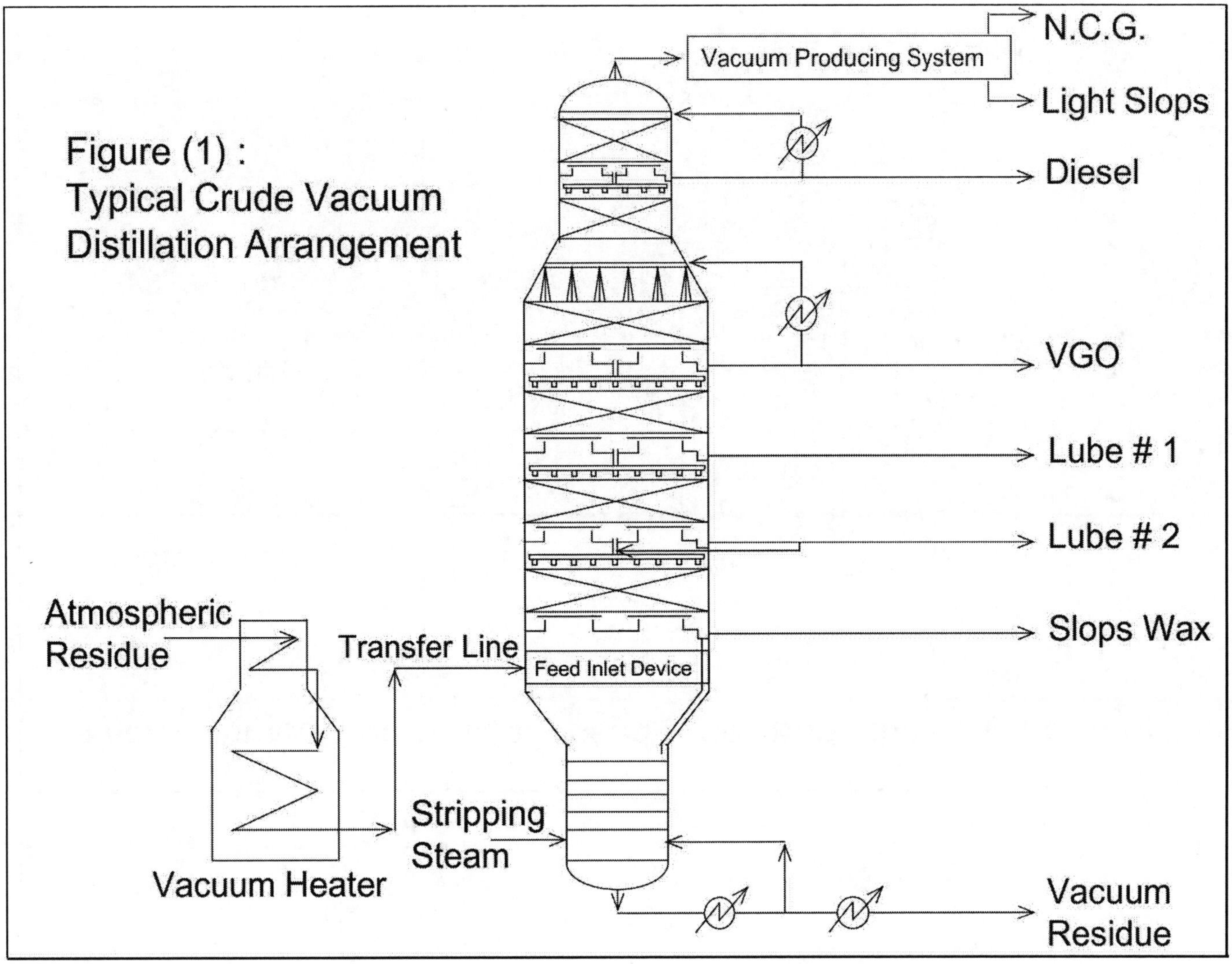

Figure (1) : Typical Crude Vacuum Distillation Arrangement

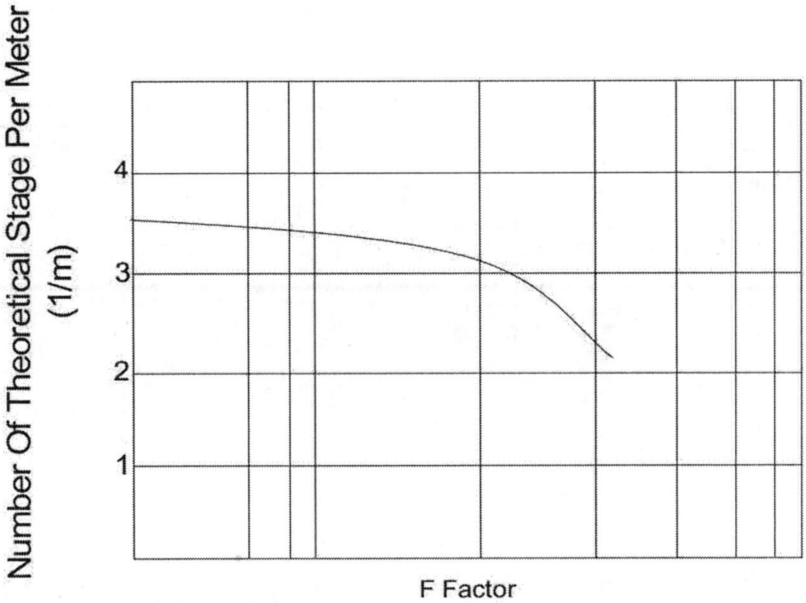

Figure (2) : Typical Packing Efficiency Curve .

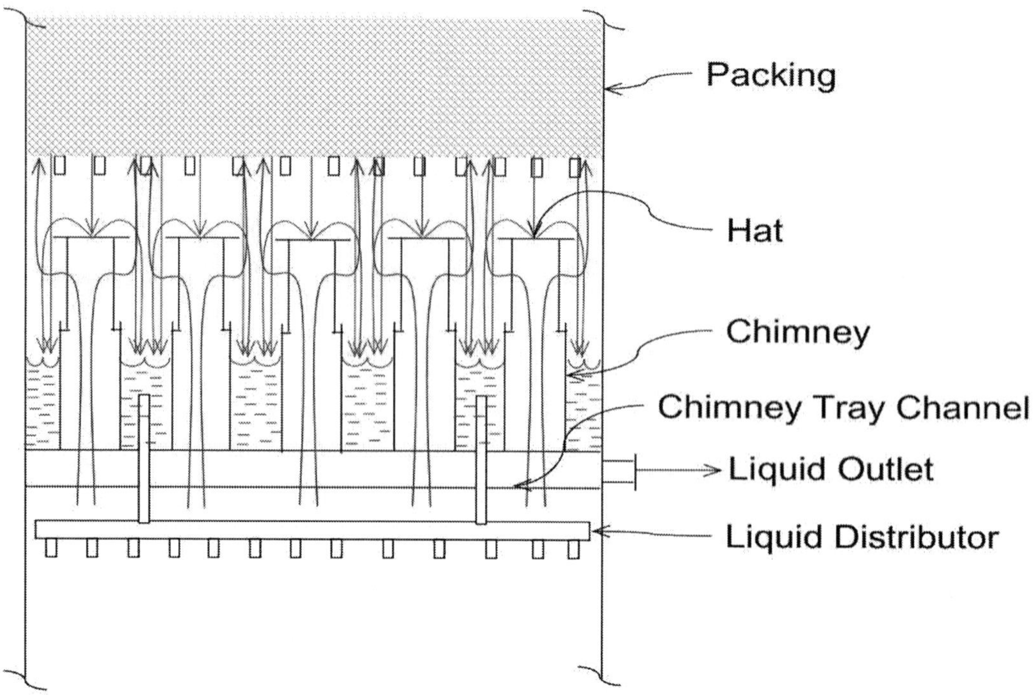

Figure (3) : Liquid and vapor passages through chimney trays .

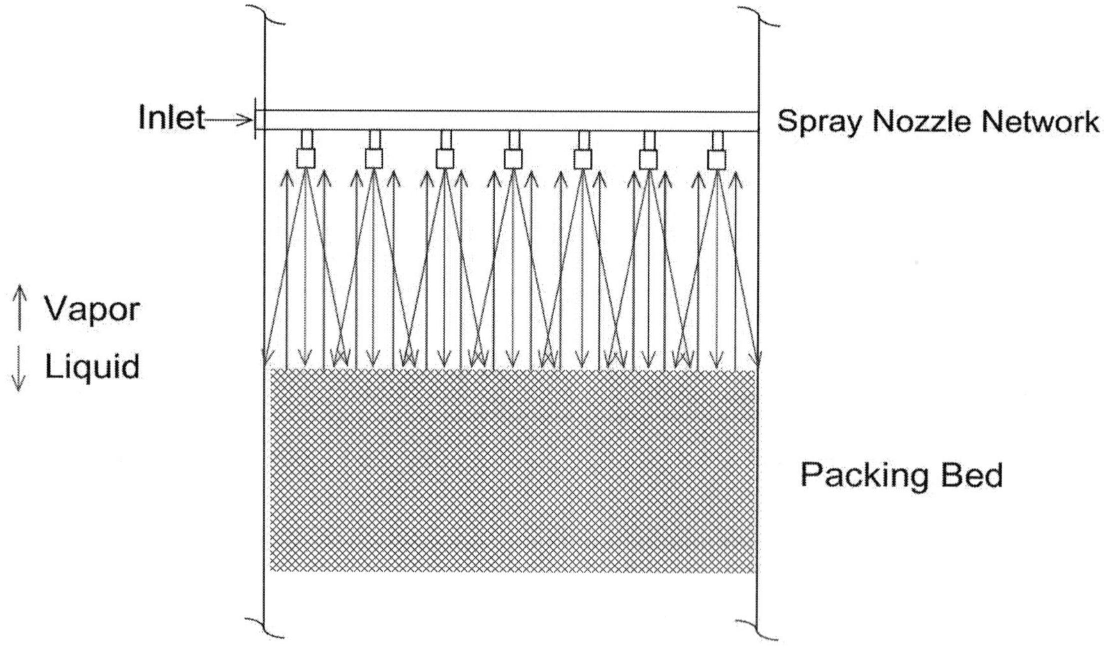

Figure (4) : Spray Nozzle Above The Packing Bed .

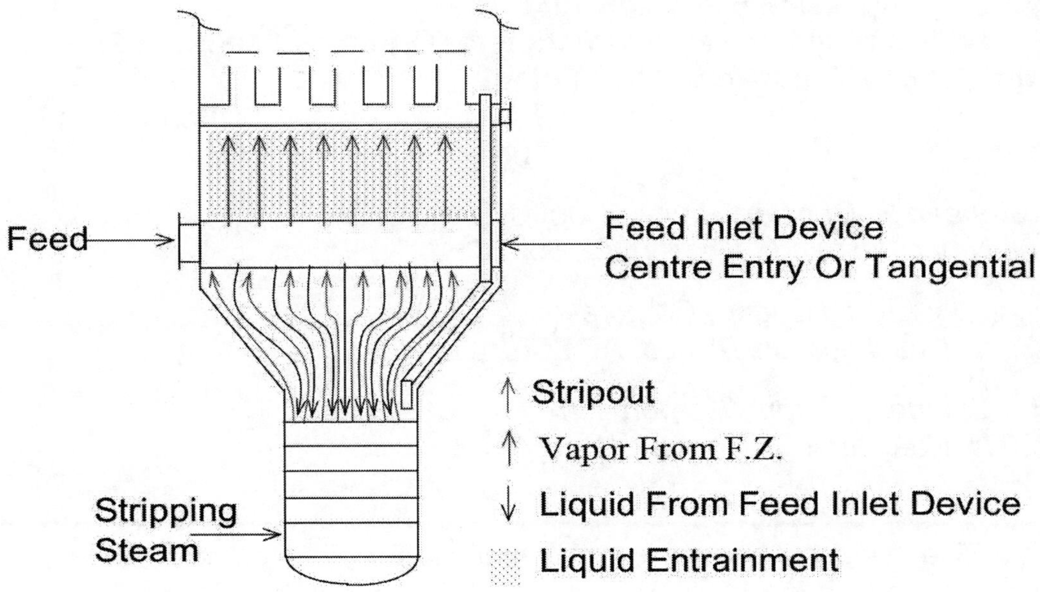

Figure (5) : Typical Flash Zone Arrangement .

(AlChe Spring Meeting Session 105f)
Reagent Promoted Solvent Extraction for Organic Chemistry: The Theory and Practice of the Next Evolution in SX Technology

By;

Dr. Reuben H. Grinstein Retired Cognis Corporation
6166 N. Via Jaspeada, Tucson, AZ 85718

Christopher Koob Cognis, Corporation
2430 N. Huachuca Dr. Tucson, AZ 85745

John Giannone, Cognis Corporation
2430 N. Huachuca Dr. Tucson, AZ 85745

The historical record is not exactly clear. The process known as solvent extraction seems to have been devised in about the eighteenth century. The first official users were probably chemists who wanted to isolate or purify chemicals they were studying. Of course there are many incidents of Analytical chemists developing solvent extraction methods to facilitate measuring various components in mixtures. Of more practical interest are the current industrial uses of solvent extraction.

There are many facets of Solvent Extraction (SX), so our objective here is to whet your taste for the possibility to solve separation problems using solvent extraction. First we will cover the basics of the solvent extraction process commonly used in industry today. Next we will explain issues needed to be resolved to design a solvent extraction process. We will show a few of the various types of equipment used in solvent extraction processes. Finally we will mention some of the commercial applications of solvent extraction.

Basically, solvent extraction is the use of a water insoluble organic material to extract some substance or group of substances from a water solution. In some cases, the extract is an unwanted byproduct or impurity to be discarded. In others, the extracted material is the desired product. In both cases, it is usually neither economically practical, nor environmentally favorable, to simply dispose of either the organic or water solution. A fully commercial solvent extraction process usually involves a second or third step to recover the solvent and to isolate the extract in some usable or disposable form.

The first applications of solvent extraction employed some solvent or mixtures of solvents to remove material from a water solution, or to solublize certain materials from solids. This process, though effective in many examples, tends not to be selective, and thus is relatively impractical for purifying or recovering many materials. Make no mistake, there are cases where the product stream contains limited compounds where one is preferentially soluable, leading to an easy extraxtion with a simple solvent.

Today, we will focus on a subset of solvent extraction; namely reagent promoted solvent extraction (RPSX). In a RPSX process, some second material, here refereed to as a reagent, is used to promote the extraction process and/or to selectively remove only certain materials from the aqueous solution. Some reagents only extract acidic materials, others remove alkaline materials, some extract by chelation effects and others remove neutral species by entrapment. We will discuss these various types of reagents later in this paper.

Before we go further, we wish to point out that are numerous examples of extraction of soluble or partially soluble material from organic streams with water or some reagent such as acid or caustic. For example, various acidic species are removed from refinery streams by caustic washing. It is not the intent of this paper to delve into this area. The application of such "neutralization" and "scrubbing" techniques is likely well know and does not fit the model discussed here.

This paper will focus on the use of RPSX to recover, purify and/or concentrate water-soluble materials. A simplified view of a potential commercially RPSX process consists of three sequences. First the water-soluble material is converted into an oil soluble species by way of the reagent. Second it is extracted into a second (oil) phase. Third the material is stripped back out of the oil phase (recovered) and the reagent/oil solution is regenerated for reuse. This process can be easily visualized in Diagram 1.

Schematic of an SX Process

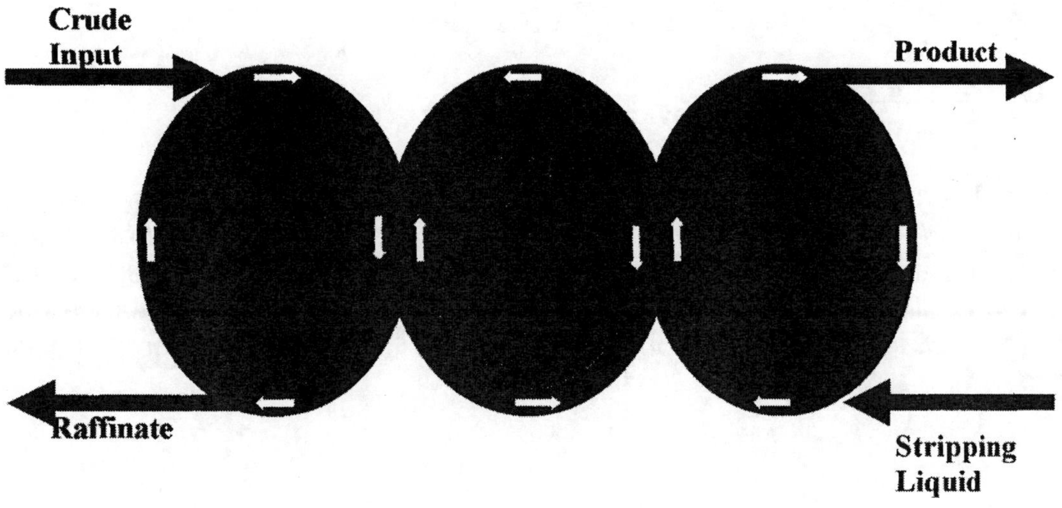

Diagram 1

In the left-hand loop, the crude water solution containing the material of interest is contacted with the reagent/solvent solution. The reagent converts the water-soluble species into an oil soluble species, which is transferred to the reagent/solvent phase. This phase is then separated from the crude in the second loop. The isolated reagent/ solvent/material solution is then contacted with some type of stripping medium which transfers the material back out of the solvent solution. The material can then be isolated from the stripping solution or used as is. In some cases, the organic solution is stripped with caustic, in others with dilute acid and in others with just hot water. For example, in the RPSX process for producing copper, the copper is stripped with dilute acid and then recovered by electrolysis. The particular process is referred to as SX-EW.

Reagents used commercially in RPSX include chelaters such as oximes, phosphates, phosphene oxides, amines, quaternary ammonium or phosphonium compounds, crown ethers and narrow range ethoxylates. Chelators such as oximes tend to be highly selective for certain metals. Certain crown ethers can also be very selective, but tend to be very expensive. Phosphates, phosphene oxides, amines and quats are less selective, though when used under carefully selected conditions, can be very economically effective. Reagents in the later group work by forming organic soluble salts with selected metal ions. This group is also used commercially for recovering organic acids, phenolics and many pharmaceuticals. A typical reaction mechanism for an amine reagent is shown in diagram

Typical Alkyl Amine Reagent

Protonation: $[R_3N]_{org} + [HA]_{aq} \leftrightarrow [R_3NH^+A^-]$

Exchange: $[R_3NH^+A^-]_{org} + [B^-]_{aq} \leftrightarrow [R_3NH^+B^-]_{org} + [A^-]_{aq}$

DIAGRAM 2

Narrow range ethoxylates (NRE), a newly introduced family of surfactant regents, work by a unique mechanism. When a solution of NRE is heated above its cloud point, it forms amorphous solids, which entrap large molecules. The solids can then be separated from the bulk of the solution. On cooling below the cloud point, the large molecules are released in a concentrated form. (see Diagram 3) This process is particularly useful in concentrating enzymes and proteins.

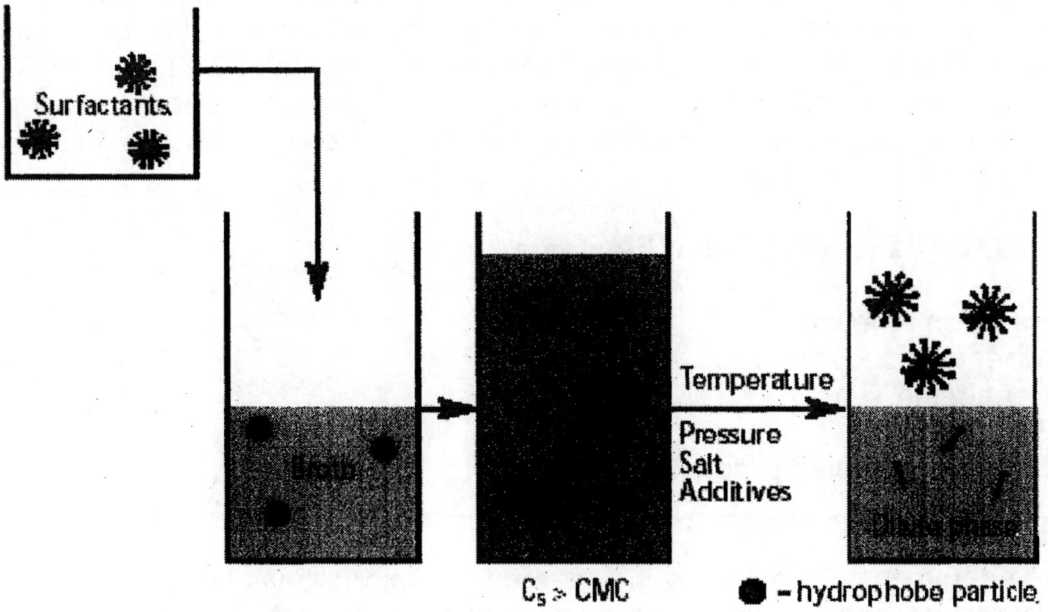

DIAGRAM 3

In developing a RPSX process, techniques similar to those used for designing distillation processes. A McCabe Thiele plot, such as the one shown below (Diagram 4), is prepared using data developed by conducting extraction measurements at different ratios of aqueous and organic.

McCabe-Thiele Diagram

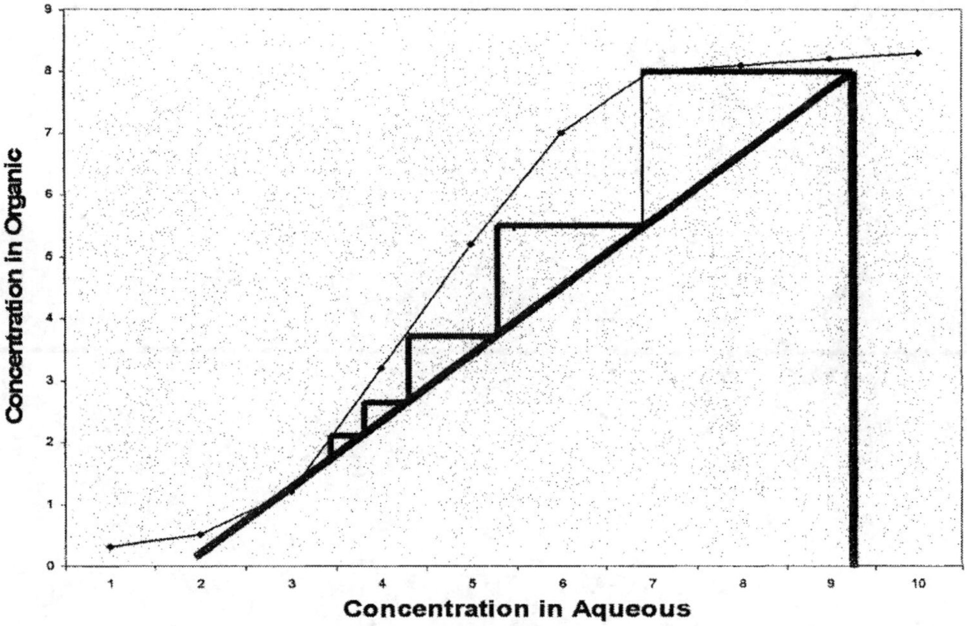

Diagram 4

The results of the analysis of the two resulting phases from each extraction are plotted on the Diagram (see the blue line). The forty five degree green line represents the desired design aqueous to organic ratio. The horizontal red lines from the green line to the blue line are used to estimate the number of extraction stages needed. There are other process specific elements that must be included in designing an efficient operation (see the list below). Each will have a profound effect on success of the process.

Table 1: Other Essential Design Elements

1. Reaction Kinetics
2. Mixer efficiency
3. Separation Time
4. Temperature
5. Form of recovered product

Equipment used commercially in RPSX includes three major types. These are mixer-settlers, counter current columns and centrifuges. Mixer-settlers are most frequently found in copper and other mining operations or processing plants. Columns and centrifuges are more often found in chemical or pharmaceutical operations. Each type of equipment has its own benefits and drawbacks. Obviously, size is a major issue. Equipment size can vary from a bench top unit such as those used to reprocess spent printed circuit board enchant to units bigger than eighteen wheelers. To illustrate the potential size of one type of unit, we show a mixer-settler (diagram 5) unit in a medium size copper operation below. Note the eighteen wheeler trailers at the bottom of the picture.

DIAGRAM 5

Hydrometalurgical production of copper from ore is probably the largest commercial application for RPSX. More than 2.5 million tons of copper is produced annually by the SX EW process. Not only does this process produce copper at roughly half the cost of traditional smelting, in many cases the copper produced is of higher quality. One mine in Arizona processes a combined volume of nearly 100,000 gallons per minute of crude copper solution through a group of mixer settlers, proving that no job is too large for RPSX.

On the small end, there are fully automated RPSX units used for reprocessing spent echant solution from printed circuit board production, which are about the size of an office desk.

Most new work is in the "middle size" application, where economics are pressed and practitioners seek higher yields at lower costs. Industrial chemical applications include purifying green phosphoric acid and recovering ferric chloride from spent pickling acid. Pharmaceutical manufacturers are producing penicillin and clevulanic acid for use in Augmentin antibiotic. Biomass fermentors use RPSX to extract citric and lactic acid from fermentation broths (see Diagram 6).

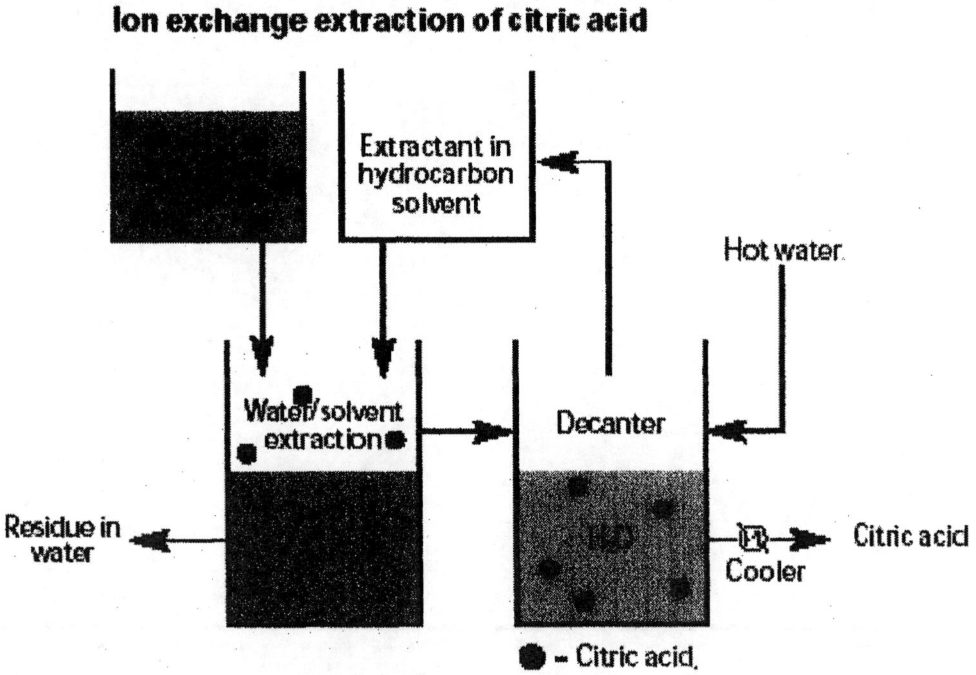

DIAGRAM 6

RPSX has even found it's way into the sustainability efforts by removing heavy metals, sulfonated surfactants and other materials from waste water. All of these application use Alkyl Amines or Alkyl Quatanary Ammonium Salts.

In conclusion, we have reviewed the concept of Reagent Promoted Solvent Extraction, given a brief description of how it works, how to design a system and mentioned three types of process equipment. Finally we have mentioned a number of commercial processes which employ RPSX. While RPSX is used in many applications, we feel there are even more opportunities.